U.S. ARMY RECONNAISSANCE & SURVEILLANCE HANDBOOK

Department of the Army

SKYHORSE PUBLISHING

All inquiries should be addressed to Skyhorse Publishing, 307 West 36th Street, 11th Floor, New York, NY 10018.

Skyhorse Publishing books may be purchased in bulk at special discounts for sales promotion, corporate gifts, fund-raising, or educational purposes. Special editions can also be created to specifications. For details, contact the Special Sales Department, Skyhorse Publishing, 307 West 36th Street, 11th Floor, New York, NY 10018 or info@skyhorsepublishing.com.

Skyhorse® and Skyhorse Publishing® are registered trademarks of Skyhorse Publishing, Inc.®, a Delaware corporation.

Visit our website at www.skyhorsepublishing.com.

10 9 8 7 6 5 4

Library of Congress Cataloging-in-Publication Data is available on file.

ISBN: 978-1-62636-099-0

Printed in the United States of America

Contents—FM 34-2-1

PREFACE vii

CHAPTER 1—INTRODUCTION 1
   Collection Management Process 3
   Solutions to Common Errors in Reconnaissance
      and Surveillance Planning 4

CHAPTER 2—RECONNAISSANCE AND SURVEILLANCE AND
INTELLIGENCE PREPARATION OF THE BATTLEFIELD 11
   Terms 11
   Reconnaissance and Surveillance Principles 13
   Intelligence Preparation of the Battlefield Process 18
   Reconnaissance and Surveillance Plan Development 36

CHAPTER 3—ASSETS AND EQUIPMENT 39
   Assets and Equipment Organic to the Maneuver Battalion 39
   Assets and Personnel Normally Supporting the Maneuver Battalion 44
   Assets and Personnel Normally Supporting the Maneuver Brigade 54

CHAPTER 4—PLANNING EFFECTIVE RECONNAISSANCE
AND SURVEILLANCE 61
   Staff Officer Responsibilities 63
   Planning 64

CHAPTER 5—METHODS OF TASKING RECONNAISSANCE
AND SURVEILLANCE ASSETS 67

CHAPTER 6—THE RECONNAISSANCE AND
SURVEILLANCE OVERLAY 71

CHAPTER 7—MONITORING THE RECONNAISSANCE AND
SURVEILLANCE EFFORT 77
   Tracking Targets and Assets 77
   Evaluating How Your Assets Report 78
   Managing Priority Intelligence Requirements 80
   Modifying the Reconnaissance and Surveillance Plan 82
   Tasking Assets 84

CHAPTER 8—AUGMENTING OR TASK ORGANIZING RECONNAISSANCE
AND SURVEILLANCE MISSIONS 87
  Task Organized with Engineers and Artillery Forward
    Observers Attached to Reconnaissance Patrol 87
  Task Organized with Signal Assets, Observation Posts, and Forward Observers
    Attached to Extended Reconnaissance Patrol 88
  Scouts with Infantry 89
  D Company, Scout Platoon, and Ground Surveillance
    Radar Effort Augmented 90

CHAPTER 9—RECONNAISSANCE AND SURVEILLANCE IN
OFFENSIVE OPERATIONS 93
  Detailed Reconnaissance 93
  Surveillance of the Objective 95
  Ongoing Reconnaissance and Surveillance Planning 95

CHAPTER 10—INTELLIGENCE SUPPORT TO
COUNTERRECONNAISSANCE 97
  Staff Officers 97
  Mission Planning 100
  Reconnaissance Fundamentals 103
  Using Intelligence Preparation of the Battlefield to Support Your
    Counterreconnaissance Effort 105
  Counterreconnaissance 108

CHAPTER 11—RECONNAISSANCE AND SURVEILLANCE IN
LOW-INTENSITY CONFLICT 115
  Factors 115
  Guerrilla/Insurgent Operations 117
  Upper Echelon Organization 118
  Equipment 123
  Forms of Guerrilla Combat 128
  Insurgent Map Symbols 133
  Movement Formations 134
  Tactical Command Basic Organization 138
  Intelligence Preparation of the Battlefield Considerations 141
  Assets 147

CHAPTER 12—ELECTRONIC WARFARE ASSET EMPLOYMENT 153
  Fundamentals 153
  Planning 155

What Type of Intelligence and Electronic Warfare
   Assets Are Needed?.. 157

APPENDIX A—MANAGEMENT TOOLS FOR RECONNAISSANCE AND
SURVEILLANCE OPERATIONS... 169

APPENDIX B—EXAMPLE OF THE RECONNAISSANCE AND
SURVEILLANCE PROCESS... 191

GLOSSARY 211

REFERENCES 219

INDEX... 221

Preface

This field manual provides tactics, techniques, and procedures (TTP) for reconnaissance and surveillance (R&S) planning, mission management, and reporting. It provides TTP for the development of intelligence to support counterreconnaissance (CR) operations. It describes employment considerations for R&S assets and defines the roles of the collection manager and maneuver brigade and battalion S2s in planning R&S operations. It describes their roles in identifying intelligence requirements to support CR operations.

This manual addresses TTP for planning and conducting R&S and developing intelligence to support CR operations at maneuver brigade and below. It can also apply to armored cavalry regiment (ACR) and separate brigades.

This manual is intended for maneuver commanders and their staffs (especially S2s), intelligence staffs and collection managers, and other personnel involved in planning and conducting R&S and developing intelligence to support CR operations. It is intended for use by both active and Reserve Components (RC) and U.S. Army Training and Doctrine Command (TRADOC) schools.

The doctrine in this publication conforms with and supports the principles contained in FM 34-1.

Unless this publication states otherwise, masculine nouns and pronouns do not refer exclusively to men.

The proponent of this publication is the United States Army Intelligence Center, Fort Huachuca, AZ. Send comments and recommendations on DA Form 2028 (Recommended Changes to Publications and Blank Forms) directly to Commander, U.S. Army Intelligence Center and School, ATTN: ATSI-TDL-D, Fort Huachuca, AZ 85613-7000.

CHAPTER 1
Introduction

Throughout history, military leaders have recognized the importance of R&S. Gaining and maintaining contact with the enemy is essential to win the battle. Our own military history contains many examples where our knowledge of the enemy, or lack of knowledge, directly led to victory or defeat.

The role of R&S has not changed on the modern battlefield; if anything, it has become even more important. Battles at the combat training centers prove that a good R&S effort is critical to successful attacks. On the other hand, a poor R&S effort almost guarantees defeat for the commander. Figure 1-1 shows attack outcome according to reconnaissance status (Blue Force [BLUFOR]). This chart was developed by the Rand Corporation in its October 1987 study, "Applying the National Training Center Experience: Tactical Reconnaissance."

RECONNAISSANCE	STATUS	BATTLE OUTCOME		
		SUCCESS	FAILURE	STANDOFF
Good	13	9	1	3
Poor	50	4	38	8

Figure 1-1. Attack outcome according to reconnaissance status (BLUFOR).

The message is clear: success on the battlefield begins with R&S and R&S begins with the intelligence officer. As the S2, you play a big role in the success or failure of your unit. But if being able to find the enemy is critical to the attack, what role does the S2 play in the defense?

Figure 1-2 shows the attack outcome according to reconnaissance status (Opposing Force [OPFOR]). This is another chart from the same Rand study. It clearly shows: if you blind the enemy, they will most likely fail in the attack. Therefore, a successful defense depends on finding, targeting, destroying, or suppressing enemy reconnaissance assets before they can report your unit's defensive positions.

RECONNAISSANCE	STATUS	BATTLE OUTCOME		
		SUCCESS	FAILURE	STANDOFF
Good	28	26	1	1
Poor	5	0	5	0

Figure 1-2. Attack outcome according to reconnaissance status (OPFOR).

This implies an aggressive CR effort that seeks out enemy reconnaissance units rather than passively screening. It also implies the coordination and active participation among the S2, S3, fire support officer (FSO), and the intelligence and electronic warfare support element (IEWSE).

This manual describes the TTP you can use to develop and execute successful R&S plans. Field Manual 34-2 and FM 34-80 contain additional information on collection management and R&S.

This is a "how to" manual. It describes how to:

■   Plan R&S operations.

■   Task R&S assets.

■   Graphically depict R&S operations.

■   Execute R&S operations.

■   Save time in the planning process.

■   Plan for intelligence support to CR missions.

■   Plan for division level assets, such as signals intelligence (SIGINT) collectors.

This manual will show you how to succeed in your reconnaissance and CR effort, giving you and your commander the best chance for victory in battle.

The intent is for you to use this manual in the field as a guide. This manual is also designed to show commanders and S3s the R&S planning process. This manual is arranged sequentially to reflect the order of the R&S and collection management processes. It will help you understand R&S a step at a time.

The better prepared you are as an S2, the better your R&S plan will be. Therefore, you should have a solid appreciation for intelligence preparation of the battlefield (IPB) and its contribution to developing an R&S plan. (See FM 34-130, Intelligence Preparation of the Battlefield, for a complete discussion of IPB.) You need to know what assets are available to you, as well as the capabilities and limitations of those assets. This supports planning and executing R&S operations.

Once you formulate your plan, you must know how to task appropriate assets. One way to disseminate the R&S plan or taskings is to develop an

R&S overlay. FM 34-80, Appendix E, describes the preparation of the R&S overlay. As you execute the plan, you should know how to monitor the R&S effort and modify the plan accordingly. To reinforce the steps in the R&S process, this manual includes examples at brigade and battalion levels of how to plan, prepare, execute, and monitor the R&S effort.

Collection Management Process

To successfully plan and execute the R&S effort, you should understand the five phases of the collection management process, and the relationship of R&S to collection management. Regardless of the echelon, you will go through the following five steps or phases to develop a collection plan and, ultimately, an R&S plan:

- Receive and analyze requirements.

- Determine resource availability and capability.

- Task resources.

- Evaluate reporting.

- Update collection planning.

Receive and Analyze Requirements

Receiving and analyzing requirements means identifying what the commander must know about the enemy, weather, and terrain to accomplish the mission. Normally, the commander's concerns are expressed as questions, termed priority intelligence requirements (PIR) or information requirements (IR).

PIR and IR are either stated by the commander or recommended by you and approved by the commander. They are the very reason R&S plans (and all collection plans) exist. You may also have requirements from higher or subordinate units; these you will prioritize and consolidate with the commander's PIR. Once you have identified all requirements, you will eventually convert them into specific items to look for.

Determine Resource Availability and Capability

In simplest terms, determining resource availability and capability means assessing what means you have to look for the specific items you have developed in the first step.

3

Task Resources

When tasking resources, you must tell a specific resource what it should look for, and how it is to report information.

At division and higher, several elements accomplish these five steps. For example, the all-source production section (ASPS) aids the collection management and dissemination (CM&D) section in analyzing requirements. The CM&D may simply task the military intelligence (MI) battalion to collect on specific requirements; the MI battalion S3 is the one who actually tasks a specific asset. In fact, very seldom does a division G2 directly task a specific asset.

At maneuver brigade and battalion levels, however, your S2 section will usually do all five steps of the collection management process. You will develop a collection plan which addresses how your unit will collect information to satisfy all intelligence requirements. Unlike division, you will normally task specific assets to collect specific information.

This essentially is the difference between a collection plan and an R&S plan: A collection plan identifies which units or agencies will collect information. An R&S plan identifies which specific assets will be tasked to collect information, and how they will do it. Therefore, as a general rule, R&S planning occurs mostly at brigade and below.

Evaluate Reporting

Is the asset accurately reporting what it sees based on its capabilities? And does the report answer the original question?

Update Collection Planning

Do you need more information to answer the question, or is it time to shift focus and begin answering another question?

Solutions to Common Errors in Reconnaissance and Surveillance Planning

This manual focuses on R&S at brigade and battalion levels. It discusses ways to improve your R&S plans and to win the battle. Many common mistakes made by S2s in the planning stage result in unproductive R&S operations. These mistakes were noted during numerous observations at the training centers and occur regularly. To avoid errors in R&S plans, use the guidelines discussed below.

Use IPB Products

Use enemy situation templates and event templates to identify areas on the battlefield where and when you expect significant events or targets to appear. These IPB products will save many hours of analysis by pin-pointing specific areas on which to focus your R&S effort.

Know Your Assets
Know the capabilities and limitations of the R&S assets available to you. This should ensure that assets are not sent on missions they are not capable of conducting, nor trained to conduct.

Provide Details
When you develop your R&S plan, provide details. Generic R&S plans do not produce the amount of information required in the time allocated.

Understand Scheme of Maneuver
It is imperative you understand your unit's scheme of maneuver before you begin to formulate the R&S plan. A well-thought-out R&S plan that does not support the scheme of maneuver is a useless effort.

Provide Guidance
Provide detailed guidance to the company and teams as they plan their patrol missions; patrols also need to coordinate with the battalion before, during, and after all missions. This should preclude useless missions and wasted lives. You should also make sure patrols have enough time to plan and execute their missions.

Know Locations of Assets
You need to continuously monitor and disseminate the current locations of friendly R&S assets. This should minimize the incidents of fratricide.

Integrate Fire Support
Involve the FSO in R&S planning so that indirect fire support is integrated into all phases of R&S operations.

Stress the Importance of R&S Missions
Subordinate elements tend to ignore collection taskings assigned by higher echelons; they sometimes consider these nonessential taskings. Make sure these subordinate element commanders understand the importance of their R&S missions. You must get the S3 or the commander involved to remedy this situation.

Participate in the Development of Mission Essential Task Lists
The battalion S2 should take an active role in the development of the scout platoon battle tasks. Figure 1-3 is an example of scout platoon

5

battle tasks. The only way to establish a proper working relationship is to train with the scout platoon leader in garrison as well as in the field. The result will be a scout platoon that understands what the S2 needs, and an S2 understanding the capabilities and limitations of the scout platoon.

BATTLE TASKS

1. Evaluate mission, formulate plans.

2. Maneuver.

3. Conduct reconnaissance operations.

4. Conduct security operations.

5. Conduct mobility and countermobility operations.

6. Operate in an NBC environment.

7. Conduct air defense measures.

8. Perform intelligence operations.

9. Perform combat support operations.

10. Sustain readiness.

CRITICAL TASKS

1. Platoon battle task: Evaluate mission, formulate plans.

    Critical Tasks:
    a) Produce a platoon fire plan.
    b) Perform precombat checks.
    c) Perform tactical planning.
    d) Employ command and control measures.
    e) Employ OPSEC.

2. Platoon battle task: Maneuver.

    Critical Tasks:
    a) Perform a tactical road march.
    b) Occupy an assembly area.
    c) Perform a passage of lines.
    d) Conduct tactical movement.
    e) Conduct a relief in place.
    f) Assist in a passage of lines.

Figure 1-3. Example of a scout platoon battle tasks list.

g) Execute actions on contact.
h) Support a hasty attack.
i) Conduct air assault operations.
j) Conduct a hasty river crossing.

3. Platoon battle task: Conduct reconnaissance operations.

    Critical Tasks:
a) Perform a route reconnaissance.
b) Perform a zone reconnaissance.
c) Perform an area reconnaissance.
d) Reconnoiter an obstacle and a bypass.
e) Execute a dismounted reconnaissance patrol.

4. Platoon battle task: Conduct security operations.

    Critical Tasks:
a) Screen a stationary force.
b) Screen a moving force.
c) Conduct dismounted security patrols.

5. Platoon battle task: Conduct mobility and countermobility operations.

    Critical Tasks:
a) Emplace and retrieve a hasty protective minefield.
b) Perform demolition guard force operations.

6. Platoon battle task: Operate in an NBC environment.

    Critical Tasks:
a) Prepare for a nuclear attack.
b) Respond to the initial effects of a nuclear attack.
c) Respond to the residual effects of a nuclear attack.
d) Prepare for a chemical attack.
e) Respond to a chemical agent attack.
f) Perform chemical decontamination.
g) Cross a chemically contaminated area.
h) Cross a radiologically contaminated area.

7. Platoon battle task: Conduct air defense measures.

Figure 1-3. Example of a scout platoon battle tasks list (continued).

```
    Critical Tasks:
a) Use passive air defense measures.
b) Take active air defense measures against hostile aircraft.

8.  Platoon battle task:  Perform intelligence operations.

    Critical Tasks:
a) Process EPWs.
b) Process captured documents and equipment.

 9.  Platoon battle task:  Perform combat support operations.

    Critical Tasks:
a) Perform resupply operations.
b) Prepare and evacuate casualties.
c) Perform platoon maintenance operations.
d) Perform field sanitation operations.

10.  Platoon battle task:  Sustain readiness.

    Critical Tasks:
a) Conduct physical training.
b) Conduct small arms and crew-served PMI and qualification.
c) Perform PMCS on vehicles and equipment.
d) Maintain property accountability.
e) Conduct battle tasks 5 to 10.
f) Counsel soldiers.
g) Update soldiers records.
```

Figure 1-3. Example of a scout platoon battle tasks list (continued).

Point Out NAI

During CR operations, point out areas throughout the battlefield where you expect enemy reconnaissance. Do not limit these named areas of interest (NAI) to just along the forward edge of the battle area (FEBA) or the line of departure (LD)/ line of contact (LC). Company and team commanders and staff must understand that CR operations extend throughout the depth of the battlefield. Enemy reconnaissance assets are trained to look deep and to conduct operations well into the rear area.

Incorporate Flexibility

Be sure to incorporate flexibility into your R&S plan. Be ready to make modifications at any time, especially in a nuclear, biological, and chemical (NBC) environment. The Combined Arms in a Nuclear/ Chemical

Environment (CANE) IIB Test explains the difficulty in collecting intelligence data in an NBC environment.

Do Not Keep the Scout Platoon Leader Waiting

Do not keep the scout platoon leader at the tactical operations center (TOC) waiting for a complete operations order (OPORD). Some results are:

■ The scout platoon deploys too late to sufficiently reconnoiter its assigned NAI.

■ Lack of time makes the scout platoon leader reluctant to exercise the initiative and flexibility necessary to ensure complete coverage. For example, scouts remain in their vehicles instead of dismounting.

■ The scout platoon fails to follow standing operating procedures (SOPS).

Do Not Overly Rely on the Scout Platoon

Commanders tend to rely too much on their scout platoons. An untrained scout platoon may display weaknesses in land navigation, selecting proper R&S positions, reporting information, and calling for indirect fire. When you do use scout platoons, try to confirm their reports with information gathered from other assets. Also, do not fall into the trap of using the scout platoon as the only R&S collection asset. When using the scout platoon, be sure the mission you give them is one they are capable of successfully completing.

These solutions apply to common problems occurring throughout the Army. Take note of these solutions and try to develop R&S plans reflecting the solutions, not the problems.

CHAPTER 2

Reconnaissance and Surveillance and Intelligence Preparation of the Battlefield

Think of developing an R&S plan as being similar to building a house. A good house needs a solid foundation. The pillars for the foundation of R&S are the actual terms used. Before going on, let us discuss some important terms.

Terms

Refer to the glossary for the definitions of reconnaissance, surveillance, and CR. Below is a discussion of these terms.

Reconnaissance

Reconnaissance is concerned with three components: enemy, weather, and terrain. You should understand that reconnaissance is active; it seeks out enemy positions, obstacles, and routes. Since movement draws attention, good reconnaissance uses stealth to avoid detection.

Surveillance

Surveillance is passive. Surveillance implies observing a specified area or areas systematically from a fixed, concealed position. A good R&S plan contains the best mix of R&S based on requirements, assets available, and the threat.

Counterreconnaissance

Essentially, CR means blinding the enemy's eyes so they cannot detect our attack, or cannot locate our defensive positions. CR missions require you to:

■ Know something about how the enemy reconnoiters.

■ Be able to locate, target, destroy, or suppress enemy reconnaissance assets.

11

Providing support to the CR mission means that you must become an expert on threat reconnaissance doctrine, tactics, unit organizations, and equipment. You must know how the enemy plans to collect information, when they do it, and with what equipment, vehicles, and organizations they collect.

Then plan how to find the enemy's reconnaissance assets before they are able to find friendly forces. You also need to understand U.S. maneuver organizations, doctrine, tactics, and capabilities, since you may be called on to provide a recommendation for organizing CR forces.

Coordinate closely with the S3 and the FSO, since much of their planning relies on your ability to predict, locate, and confirm enemy reconnaissance assets. Your knowledge of threat reconnaissance capabilities, limitations, and vulnerabilities aids the staff in developing high payoff targets (HPTs). It aids in determining how best to destroy or suppress those targets, either by lethal or nonlethal means.

At brigade level, you must get the IEWSE officer involved in CR battle planning, because using EW may be crucial to the success of the mission.

Defining R&S and CR in isolation may suggest they occur in a vacuum. Nothing could be further from the truth. R&S is a crucial phase of the intelligence cycle. As you will see, your R&S effort requires direction if it is to provide the necessary intelligence the commander needs to fight and win the battle.

You might have the impression R&S has definitive start and end points. Actually, R&S is part of a larger, continuing collection process. That process gets its direction from two things: first, the mission, and second, by extension, the IPB process.

These two things tell you:

- What to collect.

- Where to collect.

- When to collect.

- Who should collect it for you.

- Why you must collect it.

Your collection plan enables you to direct and control the collection of information. That information, once recorded, evaluated, and interpreted, becomes intelligence. Collecting information gives commanders targeting data so they can destroy enemy weapon systems and units. Your analysis can provide insight into the enemy situation to the extent that you can make an educated estimate of possible future enemy courses of action (COAs). At this point, inform your commander and the rest of the staff, then begin to develop friendly COAs for future operations.

The cycle continues endlessly. However, within the cycle you may discover, based on the picture you have developed, that you must modify the collection plan. Or, based on what you have collected, you must update the IPB terrain database.

There is an interrelationship between all aspects of the intelligence cycle. Your collection plan has a direct effect on how you:

- Process information and disseminate intelligence during the present battle.

- Direct your intelligence efforts for future battles.

The R&S plan marries the IPB with assets available for information collection. It organizes and prioritizes information requirements. This results in R&S taskings to units through the S3.

Reconnaissance and Surveillance Principles

Now that we have defined the terms, let us discuss the two principles of R&S. They are:

- Tell commanders what they need to know in time for them to act.

- Do as much as possible ahead of time.

Tell Commanders What They Need to Know
in Time for Them to Act

This principle is of paramount importance. You must develop the R&S plan so that it directly addresses what the commander wants to know. In essence, the R&S effort (as with the intelligence effort in general) is commander-oriented and commander-directed. Therefore, you cannot develop a successful R&S plan until you know exactly what the commander needs to know.

The commander's questions which positively must be answered in order to accomplish the mission are PIR. They are the start point for the R&S plan. The clearer and more precise the commander's PIR, the better you will be able to develop the R&S plan to answer them.

How do PIR come about? As part of the mission analysis process, you and your commander study the mission given to you by higher headquarters. You develop specified, implied, and essential tasks. As you do this, you should also be able to identify gaps in your understanding of the battlefield situation.

The following are examples of possible gaps:

■ Which avenue of approach (AA) will the enemy use, and when?

■ Which fixed-and rotary-wing air AAs will the enemy use?

■ How will the enemy deploy in their attack?

■ Where will the enemy commit their second echelon forces?

■ What are the enemy reconnaissance forces, tactics, and capabilities?

■ Where is the enemy main supply route (MSR)?

■ Under what circumstances will the enemy use NBC weapons? How will they be employed?

■ What are the enemy subsequent and fallback positions, and routes from their main defensive positions to the fallback positions?

■ In what strength will the enemy counterattack, and where are the armor counterattack forces?

■ Where are the enemy observation posts (OPs) and listening posts (LPs)?

■ Where are the enemy antitank (AT) helicopters?

■ Where are the enemy TOCs and tactical command posts, relays, and communications sites?

■ Where are the enemy logistic and support areas?

■ What effects of weather provide a key advantage or disadvantage to you or the enemy?

■ Where are the enemy obstacles and fire sacks?

■ Where are the enemy main defensive positions?

■ Where and when will the enemy counterattack, and how will they counterattack?

■ Where are the enemy AT weapons?

■ Where is the enemy artillery?

Essentially, you and your commander try to zero in on exactly what will determine the outcome of the battle. Many times, the commander will tell you what is of the most concern. It is the responsibility of the commander to state PIR. However, the S2 and the S3 can assist in this process by presenting their own analyses of the mission.

Remember, PIR drive your R&S efforts so it is critical that you understand just exactly what your commander needs to know in order to fight.

Try to keep the commander's PIR as specific as possible. The more general the question, the harder it is to answer. Instead of asking, "With what force will the 34th Motorized Rifle Regiment attack?", try to discern exactly what it is your commander is looking for. If the commander wants to know how the 34th will initially deploy, it is better to ask, "Will the 34th attack in column, with two battalions leading, or with three battalions on line?"

Similarly, is the commander concerned with finding the enemy's main defensive area, or is he really worried about locating company and platoon positions within the main defensive area? How you phrase the PIR has a direct bearing on how you answer those PIR.

To better focus R&S efforts, keep PIR down to a manageable number. Normally, you will only be able to concentrate on three or four at any one time. Of course, the mission and the commander's needs may sometimes dictate more. Having a large number of priorities defeats the purpose of having PIR in the first place.

Other questions the commander may have regarding the enemy, weather, and terrain of a lesser priority than PIR are called IR. Examples of IR are:

■ Where or what is the enemy's immediate objective?

■ Will the enemy employ smoke?

■ Where are the usable river fording points?

Answers to these questions may not be critical to the immediate success of the mission, but they will certainly help provide answers to those critical questions. Keep in mind that IR may at some point become PIR, and vice versa, depending on the situation.

Once developed, PIR should be disseminated to subordinate, adjacent, and higher units. In this way, you inform everyone of the questions you need answered. Remember, although you may not be able to answer a specific PIR at your level, our higher headquarters may be able to. Disseminating your PIR also tells everyone what you are most interested in.

Use the intelligence annex and intelligence summaries to disseminate PIR. Your PIR and IR also determine your request for intelligence information (RII). The RII is the best way to let your higher headquarters know your information needs. Your higher headquarters does not always know what specific information you may need.

Your commanders PIR give you a direction in which to start your R&S planning. Subsequently, most of your time will be spent doing detailed planning and analysis, all focused on answering those PIR. Essentially, you will study the effects of enemy, weather, and terrain on the battlefield and the mission.

The best way to study the enemy, weather, and terrain is through the IPB process. IPB:

- Enables you to focus analytical efforts on a specific part of the battlefield.

- Gives you a way to systematically examine the terrain and weather effects on enemy and friendly actions.

- Helps you determine the effects on R&S activities.

- Gives you an in-depth view of how the enemy fight, how they reconnoiter, and where they may be most vulnerable.

Most importantly, IPB gives you a way to synchronize your R&S plan with the general battle plan. Figure 2-1 illustrates the commander's decision-making process (supported by IPB). If it is done right, IPB is people-intensive and time-consuming. This brings us to a second principle of R&S.

Do as Much as Possible Ahead of Time
The first four functions of the IPB process are homework functions. That means you build a database on terrain and weather conditions and enemy order of battle (OB) before hostilities. For example, your

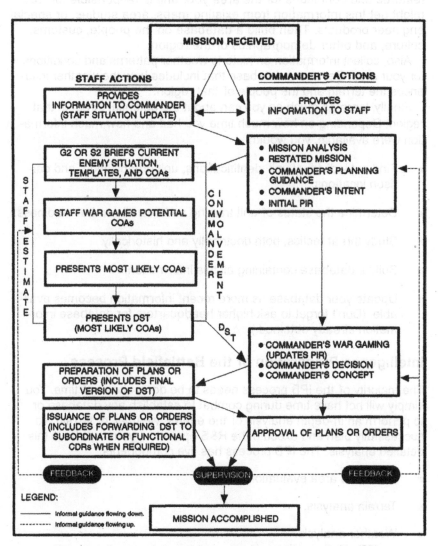

Figure 2-1. Commander's decision-making process (supported by IPB).

unit has received a new contingency mission for some part of the world. Given this mission, begin collecting information on terrain features and conditions for the area your unit is responsible for. You might get this information from existing maps, area studies, or special engineer products. Then build a database on the people, customs, culture, and other demographics of the region.

Also, collect information on historic weather patterns and conditions for your area. Develop a database that includes how the weather influences the terrain and the people of that region.

Finally, learn as much as you can about the potential threat in that region. Depending on how much time you had and how much information were available, you would:

- Find out individual unit identifications, unit organization, and garrison locations.

- Determine the status of unit training and overall combat readiness.

- Study threat tactics, both doctrinally and historically.

- Build a database containing all this information.

- Update your database as more recent information becomes available. (Don't forget to ask higher headquarters for database information already obtained.)

Intelligence Preparation of the Battlefield Process

The majority of the IPB process needs to be done ahead of time. You simply will not have time during combat to establish any database or to perform an in-depth analysis of the enemy, weather, or terrain. To successfully plan and execute the R&S operation, you must have this detailed analysis. The IPB process has five components:

- Battlefield area evaluation (BAE).

- Terrain analysis.

- Weather analysis.

- Threat evaluation.

- Threat integration.

Refer to FM 34-130, Intelligence Preparation of the Battlefield, for detailed information on IPB.

Battlefield Area Evaluation

BAE is the first step of the IPB process. Begin your analysis by figuring out what part of the battlefield should be of interest to you and your commander. The end result of this step is the identification of the area of interest (AI): that part of the battlefield which contains significant terrain features or enemy units and weapon systems that may affect your unit's near or future battle.

BAE is a crucial step in the IPB process because it focuses your analytical efforts on a finite piece of the battlefield. By extension, it will also provide geographic limits to your R&S and collection efforts.

The commander bases the unit's AI on many things. It is normally an expansion of your unit's area of operations (AO). It should be large enough to provide answers to the commander's PIR, yet small enough to prevent your analytical efforts from becoming unfocused. Determining the AI depends on the unit mission and threat capabilities. For example, if your unit is to attack, your AI should extend across your LD/LC up to and surrounding your intermediate and subsequent objectives.

If the mission is to defend, the AI should extend far enough to include any possible units that might reinforce against you. You can base your AI considerations in terms of time and on how fast you or the enemy moves. Figure 2-2 lists general distance guidelines in hours and kilometers; use this to determine your unit's AI.

COMMAND ECHELON	AREA OF INTEREST	
	Hours	Kilometers
Battalion	Up to 12	Up to 15
Brigade	Up to 24	Up to 30
Division	Up to 72	Up to 100

Figure 2-2. General distance guidelines.

Considerations for your AI should be expressed in terms of distance, based on:

- How your unit attacks.

- How the enemy attacks.

- What your commander needs to know.

For example, a battalion commander fighting an attacking enemy using Soviet tactics is normally interested in 1st- and 2nd-echelon battalions of 1st-echelon regiments.

Doctrinally, these units would normally be from 1 to 15 kilometers from our FLOT. Therefore, the AI should extend forward at least 15 kilometers.

You must determine your AI during mission analysis. Your commander and S3 play a big part in formulating the AI. They tell you what their intelligence concerns are. Like PIR, your unit's AI must be the commanders and must be sent to higher headquarters. Figures 2-3 and 2-4 show examples of AIs for defensive and offensive missions. Figure 2-5 shows both defensive and offensive. It will help in determining your unit's AI.

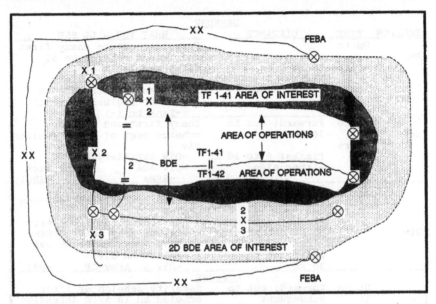

Figure 2-3. Area of interest in the defense.

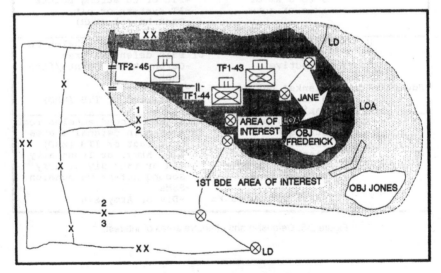

Figure 2-4. Area of interest in the offense.

DEFENSE			
ECHELON	TIME	DISTANCE	WHAT YOU WILL SEE

Bn	Up to 12 hours	15 km
		
		Flanks: 3 to 6 km either side
			Forward: out to enemy first and second echelon bn of first echelon regts.
			  -Regimental artillery
			  -Regimental air defense
			  -Div or regimental recon
			    -Tank bn (-) (MRR)

Bde	Up to 24 hours	Forward: out to 30 km
		
		Flanks: 6 to 10 km either side
			Enemy first and second echelon regt of first echelon div
			  -Div artillery -ITB
			  -Div ADA -tank regt
			    -SSMs (OF MRDs)

Div	Up to 72 hours	Forward: out to 100 km
		
		Flanks: 2 to 30 km either side
		
		Rear: out to 30 km
			Enemy first and second echelon div of first echelon armies
			  -Div, Army, or front arty
			  -Div or Army ADA -SSMs
			  -Div, Army, or front avn
			  -Follow-on div
			    -Div or Army CP -OMG

OFFENSE			

Bn	Up to 12 hours	Forward: out to subsequent objective
		Flanks: out to 3 to 6 km of axis or zone
			Security echelon or first echelon bn in main defensive area
			  -Div or regt recon
			  -Plt or co strong points
			  -AT systems or plt
			  -Tank bn (-) (MRR)

Bde	Up to 24 hours	Forward: out to subsequent objective
		
		Flanks: out to 6 to 10 km of axis or zone
			First and second echelon bn of main defensive area
			  -Co/bn strong points/fire sacks
			  -AT battery
			  -Tank regt or ITB (MRD)
			  -AT bn

Div	Up to 72 hours	Forward: out to subsequent objective
		Flanks: out to 25 to 30 km of axis or zone
		
		Rear: out to 30 km
			First and second echelon regt in Army main defensive area
			  -Tank regt or ITB (MRD)
			  -Div, Army, or front arty
			  -ITR or tank div of Army Second defensive echelon
			  -SSMs
			  -Div or Army avn

Figure 2-5. Defensive and offensive areas of interest.

22

Terrain Analysis and Weather Analysis

The next two steps in the IPB process are terrain and weather analyses. Essentially, these are detailed studies of how the terrain and weather will affect both friendly and enemy operations. Specifically, terrain and weather will dictate how effective R&S assets will be, and where they should go to be most effective. Your knowledge of terrain and weather will allow you to anticipate effects on friendly and enemy R&S systems and operations.

Terrain analysis and weather analysis should start as soon as you have determined your AI. Do not wait until you deploy to start your analysis! The more prepared you are, the better the R&S plan will be. Figures 2-6 and 2-7 show specific uses and effects for terrain and weather analyses.

TERRAIN FACTOR	R&S APPLICATION
Observation	Ensuring LOS for GSRs Emplacement of OPs/LPs and NODs Ensuring radio LOS for comm with R&S assets Templating enemy R&S asset locations Templating enemy smoke and obscurant employment
Field of Fire	Templating enemy defensive positions Templating possible enemy obstacle locations Coordinating friendly supporting fires for patrols and other R&S assets
Concealment and cover	Ensuring routes for friendly patrols and scouts Emplacement of REMBASS sensor strings Emplacement of LPs, OPs, and GSRs Templating enemy patrols Templating enemy obstacles
Obstacles	Ensuring routes for friendly patrols and scouts Emplacement of REMBASS sensor strings Emplacement of LPs, OPs, and GSRs Templating enemy patrols Templating enemy obstacles
Key terrain	Emplacement of LPs, OPs, GSRs, NODs, and REMBASS Ensuring routes for friendly patrols or scouts Templating enemy patrols, recon, or obstacles Templating enemy movement or defensive positions
AAs	Emplacement of OPs, LPs, GSRs, NODs, and REMBASS Templating enemy movement or defensive positions Templating enemy recon effort

Figure 2-6. Special uses and effects of terrain.

24

ENVIRON-MENTAL EFFECTS	PVS5	PVS2	PVS4	DRAGON THERMAL	TVS2/5	TVS4	TOW THERMAL PAS7/UAS11	GSR	REMBASS
Reduced visibility (in meters) (darkness, smoke and fog)	MLT 200	MLT 400	MLT 600	MLT 1,000	MLT 1,200	MLT 2,000	MLT 3,200	No EFX	No EFX
Surface winds (in knots)	No EFX	No EFX	No EFX	No EFX	No EFX	No EFX	No EFX	SGT 20 MGT 7	SGT 45 MGT 15
Temperature (in fahrenheit)	SGT 125 MLT 32	SGT 125	SGT 125 MLT -20	No EFX	SGT 125 MLT (5) -20	SGT 125	SLT (PAS7) -25 MLT (UAS11) 20	No EFX	No EFX
Precipitation	All sensors are severely degraded by heavy rain or snow.								

LEGEND: S = Severe degradation M = Moderate degradation LT = Less than GT = Greater than EFX = Effects

Figure 2-7. Effects of environment on R&S.

Threat Evaluation

Once you have analyzed terrain and weather, begin a thorough study of enemy:

■ Composition.

■ Disposition.

■ Tactics.

■ Training.

■ Combat readiness.

■ Logistic support.

■ Electronic technical data.

■ Personalities.

■ Other miscellaneous factors.

This study results in threat evaluation, the fourth step in the IPB process. During this step:

25

■ Develop a doctrinal template file.

■ Build up your threat database.

■ Evaluate threat capabilities.

Doctrinal templates are important because they show how the enemy doctrinally attacks or defends in various situations. Knowing how the enemy defends will tell you what you ought to look for in order to confirm that they are, in fact, defending.

Knowing how the enemy employs reconnaissance in the attack will help you target them, allowing you to destroy or neutralize those assets. It also helps you determine which of those assets are most important to the enemy's reconnaissance effort.

Figures 2-8 and 2-9 are examples of doctrinal templates you might use specifically for R&S planning. Whenever you use doctrinal templates, you must temper them with some reality. For example: a Soviet regimental attack template has set doctrinal sector widths. It serves no purpose to place this over a map where a battalion falls outside an AA. There is enough leeway, even in Soviet doctrine, to conform to terrain limitations; when using the template you must make those same allowances.

A careful study of threat doctrine tells how fast they will attack in various situations. This information will become very important later on. For right now, remember during threat evaluation that you determine enemy doctrinal rates of advance. Figure 2-10 is a table of enemy rates of advance for specific situations and terrain.

Finally, knowing how the threat uses weapon systems and units gives you an appreciation of which are most important to the enemy commander in a particular situation. These important weapon systems and units are called high value targets (HVTs). They are the starting point for the target value analysis process. Target acquisition is an important aspect of R&S and CR. Target value analysis will play a big role in your R&S planning.

Threat Integration

Remember, the four IPB steps should be started before deployment. They ought to be part of your day-by-day intelligence operations. You are now at the point where you can pull together what you have developed about the enemy, weather, and terrain and apply it to a specific battlefield situation.

This step is threat integration. You will discover you can also perform some threat integration functions ahead of time. The first such function is to develop a series of situation templates depicting how you think the enemy will deploy assets.

26

Situation Template

The situation template takes what is on the doctrinal template and integrates what you know about weather and terrain. The situation templates will show how an enemy unit might modify its doctrine and tactics because of the effects of weather and terrain.

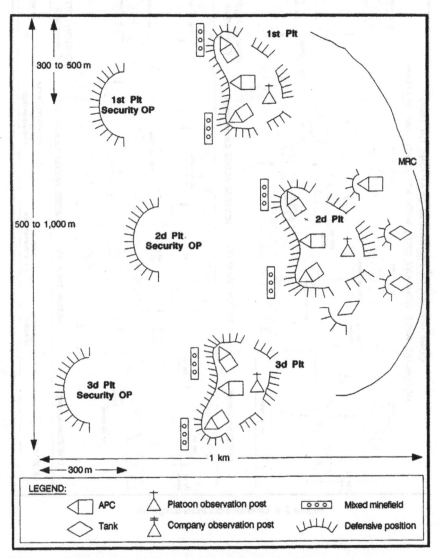

Figure 2-8. Doctrinal template of an MRC (reinforced) strong point.

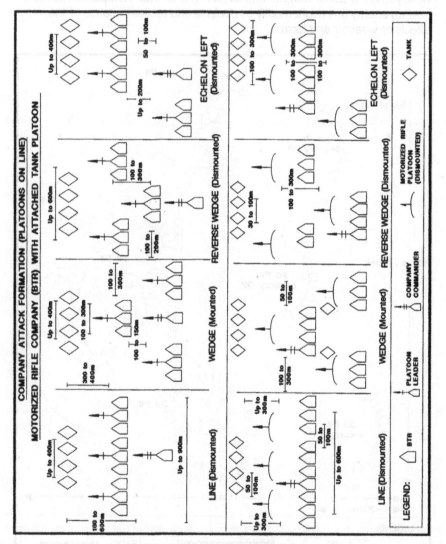

Figure 2-9. Offensive doctrinal tempelate.

RATES OF MARCH

UNOPPOSED TACTICAL THREAT RATES OF MARCH

Unpaved road or trail	20 to 30 km/h	1 km/h 2 min
Cross country	5 to 15 km/h	1 km/h 4 min
Paved roads	30 to 40 km/h	6 km/h 10 min
Assault	11 to 22 Km/h	1 km/h 3 min

RATES OF MOVEMENT (km/h)

RATIO THREAT : OPPOSITION	PREPARED DEFENSE			HASTY DEFENSE		
	GO	SLOW GO	NO GO	GO	SLOW GO	NO GO
Intense resistance 1 : 1	.6	.5	.3	1.0	.8	.4
Very heavy resistance 2 : 1	.9	.6	.4	1.5	1.0	.6
Heavy resistance 3 : 1	1.2	.75	.5	2.0	1.3	.8
Medium resistance 4 : 1	1.4	1.0	.5	2.4	1.75	.9
Light resistance 5 : 1	1.5	1.1	.6	2.6	2.0	1.0
Negligible resistance 4 : 1	1.7+	1.3+	.6 +	3.0+	2.3+	1.1+

Opposed doctrinal rates of advance

6 km/h covering force	1 km/h 10 min
2 km/h MBA	1 km/h 30 min
5 km/h rear	1 km/h 12 min

Figure 2-10. Threat rates of advance.

Figure 2-11 is an example of one situation template. It is important to understand that you should develop as many situation templates as there are enemy COAs. This allows you to thoroughly examine what options the enemy has for each COA.

For example, you may discover enemy forces have to use specific bridges, road intersections, or mobility corridors (MCs) for a specific

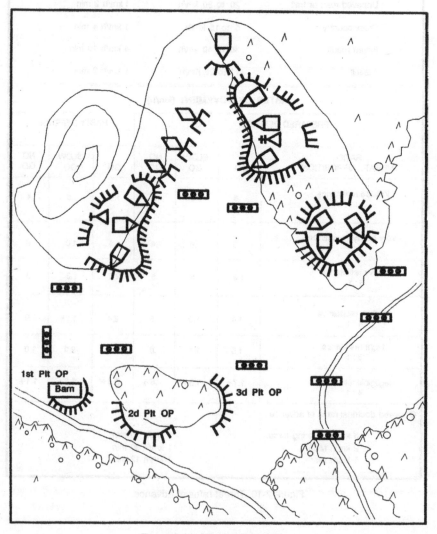

Figure 2-11. Situation template.

COA. Or you may discover that the terrain offers the enemy several choices to attack. Or you may determine the terrain offers a limited number of suitable enemy defensive positions. And you may learn that the terrain only provides a limited number of concealed routes for enemy reconnaissance to enter your sector.

The bridges, road intersections, and possible defensive positions you have identified become NAI. Focus your attention on these NAI because it is there you expect something to happen. What you see or fail to see at your NAI will confirm whether or not the enemy is doing what you expected them to do, as projected on the situation template. NAI do several things for you. They:

■ Focus the collection effort on specific points or areas of the battlefield.

■ Tell you what to look for and when you should expect to see it, at those points or areas on the battlefield (based on the situation templates).

■ Enable you to decide which of your R&S assets are best suited to cover a particular NAI. For example, a point NAI for a road intersection where you expect to see a motorized rifle company (MRC) might require seismic-acoustic sensors. On the other hand, an area NAI designated for a dug-in BMP platoon might require a dismounted patrol.

Remember, one of the things you did during threat evaluation was to determine enemy rates of advance. You now put this knowledge to work by developing time phase lines (TPLs). Think of TPLs as snapshots of an enemy or a friendly frontline trace. A series of TPLs would portray friendly or enemy movement over a period of time.

Event Template

If you combine NAI with TPLs, you will be able to show approximately when and where you would expect to see enemy critical events occur. This is basically what the event template does. Figure 2-12 is a sample event template.

The event template allows you to:

■ Confirm or deny your situation templates.

■ Gauge enemy and friendly rates of movement.

■ Compare rates of movement between MCs and AAs.

■ Cue other collection assets based on friendly and enemy movement.

Of all IPB products, the event template is the most important product for the R&S effort. As you will see, the event template is also the basis for the decision support template (DST).

In many situations you might find it helpful to calculate how long an enemy unit would take to move from one NAI to another. Normally, your

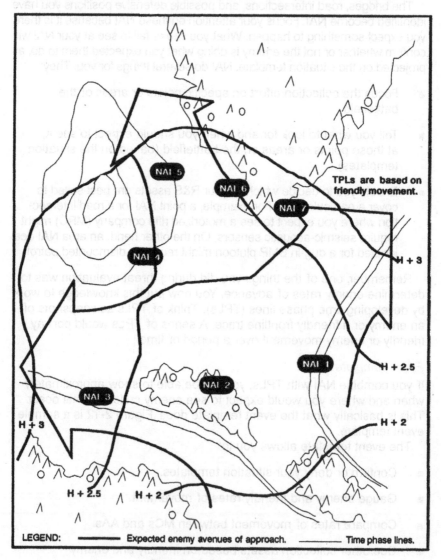

Figure 2-12. Event template.

calculations will be based on opposition and doctrinal rates of advance. Situational aspects such as weather, terrain, and your previous hindering actions are also factored in.

Comparing actual movement rates with your calculations will tell you whether the enemy is moving slower or faster than expected. It will also help you predict how long it will actually take the enemy to reach a certain point (your FEBA, for example).

Event Analysis Matrix

The event analysis matrix is a tool used along with the event template to analyze specific events. Figure 2-13 shows examples of event analysis matrixes.

Basically, you calculate the not earlier than (NET) and the not later than (NLT) times lead elements of a unit will arrive at an NAI. Determine the distance between NAI and multiply the distance by the expected rate of advance.

For example, suppose the distance between NAI 1 and NAI 2 is 2.5 kilometers. Suppose also, for the sake of this example, that the enemy expected rate of advance is 6 kilometers per hour, or 1 kilometer every 10 minutes. Use this formula to calculate time:

$$distance/rate = time$$

2.5 km ÷ .1 km (1 km every 10 minutes) = 25 minutes. Therefore, it takes the unit 25 minutes to travel from NAI 1 to NAI 2.

Decision Support Template

The final IPB product is the DST. The purpose of the DST is to synchronize all battlefield operating systems (BOS) to the commander's best advantage. The DST consists of target areas of interest (TAI), decision points or lines, TPLs, and a synchronization matrix. Figure 2-14 shows a DST.

There are many important things you should know about the DST. First, the DST is a total staff product, not something the S2 makes in isolation. Although you may begin the process of developing the DST, the S3 and the commander drive the development.

Second, the DST is a product of war gaming. Together with the rest of the staff, you develop friendly COAs which consider what you envision the enemy doing. As a result of this action, reaction, and counteraction war game, you identify actions and decisions that may occur during the battle.

Third, the R&S plan must support the DST.

AVENUE OF APPROACH:			COORDINATES:		
MOBILITY CORRIDOR: 1			COORDINATES: PA 2071 - PA 1076		
NAI ACTIVITY	DISTANCE (KM)	RATE OF MOVEMENT	NET NLT		TIME OBSERVED TIME CONFIRMED
1. Enemy recon elements, groups, and patrols move along MC 1.	—	—	— H-hours		
2. Activity is as per NAI 1.	18 km	20 km	H + 54 min H + 69 min		
3. Recon elements assume screen positions.	20 km	20 km	H + 114 min H + 144 min		

AVENUE OF APPROACH:			COORDINATES:		
MOBILITY CORRIDOR: 1			COORDINATES: PA 4850 - PA 2840		
NAI ACTIVITY	DISTANCE (KM)	RATE OF MOVEMENT	NET NLT		TIME OBSERVED TIME CONFIRMED
1. Enemy MRBs conduct night river crossing with illumination.	—	—	— H-hours		
2. Enemy counterattack with reinforced TB.	20 km	20 KmPh	H + 1 H + 75 min		
3. Enemy reinforce MRB shifting into prebattle formation.	18 km	20 KmPh	H + 104 min H + 144 min		

Figure 2-13. Event analysis matrixes.

Fourth, you can use the DST, as well as the general battle plan, to synchronize the R&S effort.

As a result of the war-gaming process, the staff identifies HPTs—those enemy weapon systems and units that must be acquired and successfully attacked for the success of the friendly commander's mission.

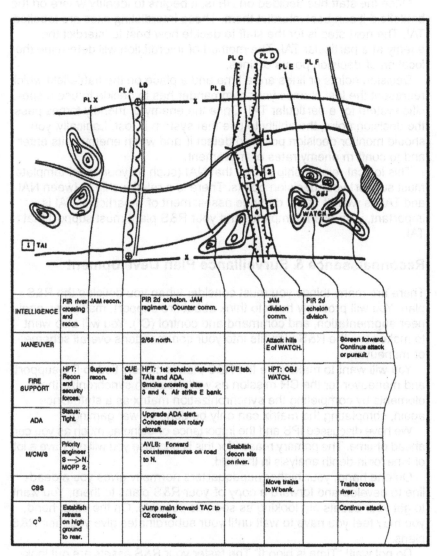

	PIR river crossing and recon.	JAM recon.		PIR 2d echelon. JAM regiment. Counter comm.		JAM division comm.	PIR 2d echelon division.
INTELLIGENCE	PIR river crossing and recon.	JAM recon.		PIR 2d echelon. JAM regiment. Counter comm.		JAM division comm.	PIR 2d echelon division.
MANEUVER				2/68 north.		Attack hills E of WATCH.	Screen forward. Continue attack or pursuit.
FIRE SUPPORT	HPT: Recon and security forces.	Suppress recon.	CUE tab	HPT: 1st echelon defensive TAIs and ADA. Smoke crossing sites 3 and 4. Air strike E bank.	CUE tab.	HPT: OBJ WATCH.	
ADA	Status: Tight.			Upgrade ADA alert. Concentrate on rotary aircraft.			
M/CM/S	Priority engineer S --->N. MOPP 2.			AVLB: Forward countermeasures on road to N.		Establish decon site on river.	
CSS						Move trains to W bank.	Trains cross river.
C³	Establish retrans on high ground to rear.			Jump main forward TAC to C2 crossing.			Continue attack.

Figure 2-14. Decision support template.

35

The staff identifies HPTs from the list of HVTs you developed during threat evaluation. (See FM 6-20-10, TTP for the Targeting Process.)

Once the staff has decided on HPTs, it begins to identify where on the battlefield it can best interdict them. These interdicting sites are labelled TAI. The next step is for the staff to decide how best to interdict the enemy at a particular TAI. The method of interdiction will determine the location of decision points or lines.

Decision points or lines are a time and a place on the battlefield which represent the last chance your commander has to decide to use a specific system for a particular TAI. Once the enemy or friendly forces pass the decision point, the ability to use that system is lost. Logically, you should monitor decision points to detect if and when enemy units enter and to confirm enemy rates of movement.

This logical relationship shows that NAI (such as your event template) must support your decision points. There is a relationship between NAI and TAI as well. If battle damage assessment of a particular TAI is important, your event template (and your R&S plan) must support that TAI.

Reconnaissance & Surveillance Plan Development

There are many things you must consider when you develop the R&S plan. You will probably have to think about fire support, maneuver, engineer augmentation, and command and control (C^2). You will also want to make sure the R&S plan fits into your commander's overall scheme of maneuver.

You will want to make sure the R&S plan is closely tied to fire support and maneuver for the CR mission as well. You can synchronize these elements by completing the synchronization matrix as a staff. Once again, completing the matrix can only be done by war gaming.

We have discussed IPB and the importance of doing as much as you can ahead of time. The primary reason for this is because you will not have a lot of time for in-depth analysis in the field.

On one hand, your higher headquarters normally gives you a deadline to develop and forward a copy of your R&S plans to them. You want to get your assets out looking as soon as possible. On the other hand, you may feel you have to wait until your subordinates give you their R&S plans.

Do not wait! "Time is blood!" The faster your R&S assets are out looking, the more time they will have to find what you want.

Do not delay your planning because you do not have a complete situation template, or because you do not have all your subordinate units' plans. If necessary, give your R&S assets an initial mission and update the mission when you have had the time to do more detailed planning.

The technique that allows you, your subordinate units, and your higher headquarters to conduct R&S planning simultaneously is the use of limits of responsibility. A limit of responsibility is a boundary defining where a particular unit should concentrate its R&S efforts.

In essence, a limit of responsibility is a "no further than" line; it tells the unit, "your R&S responsibility stops here." Figure 2-15 is an example of limits of responsibility for battalions, brigades, and divisions. They may be tied to a unit's AI or may depend entirely on mission, enemy, terrain, time, and troops available (METT-T).

The key is that limits of responsibility allow each echelon to formulate its R&S plan independently. It is important to note that an R&S plan is never a finished product. Because the situation and the operational plan will most likely change, the R&S plan must change to fit them.

This chapter discussed R&S and CR definitions, PIR, the contributions of IPB to R&S, and limits of responsibility. If you understand these concepts, you have a solid foundation on which to build your R&S plan. The next chapter talks about assets you may have available to you. These will be the bricks for you to actually build your plan.

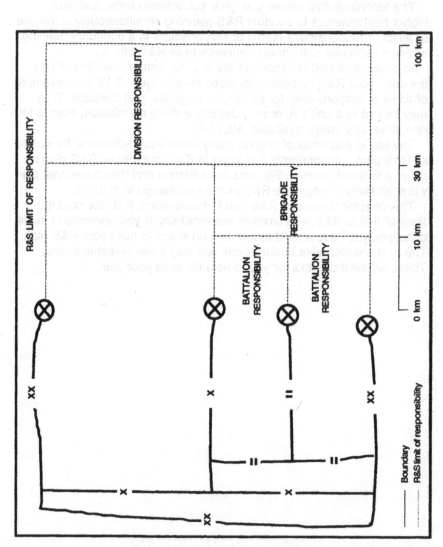

Figure 2-15. R&S limits of responsibility.

CHAPTER 3
Assets and Equipment

Before you can develop an R&S plan you must know the characteristics of available assets and equipment. This chapter discusses the organic and supporting R&S assets and equipment available to you, the maneuver battalion, and the brigade S2. This chapter also discusses the basic capabilities and limitations of these assets.

Due to security classifications, detailed information on some assets is not included. Refer to the appropriate field manuals for further details. Some assets are staff officers; this chapter gives you the types of information they can provide and equipment they might use.

Assets and Equipment Organic to the Maneuver Battalion

At the battalion level the commander is fighting close-in operations. To support the commander, focus the collection effort at the 1st- and 2nd-echelon battalions of 1st-echelon regiments (out to 15 kilometers). The assets available to conduct R&S missions at the battalion are limited. Since the assets available to conduct R&S missions at the battalion are limited, the S2 faces a challenge when planning R&S operations. Some of the available assets are scouts, patrols, OPs/LPs, soldiers, and equipment.

Scout Platoon

The scout platoon's primary missions in support of the battalion are reconnaissance and screening. FM 17-98 contains a detailed discussion of scout platoons. Presently all scout platoons are being reequipped with 10 high mobility multipurpose wheeled vehicles (HMMWVs).

Capabilities
The scout platoon can perform missions:

■ Mounted or dismounted.

■ In various terrain conditions.

■ Under all weather conditions.

■ Day and night.

39

The primary mission of the scout platoon is reconnaissance. The scout platoon, with fire support protection, can conduct reconnaissance missions 10 to 15 kilometers beyond the FEBA. These distances vary with the type of scout platoon and METT-T.

The scout platoon is the only asset found at the maneuver battalion specifically trained to conduct reconnaissance. Other missions the scout platoon conducts are:

- Quartering party duties.

- NBC reconnaissance, including chemical detection and radiological surveying and monitoring.

- Limited pioneer and demolition.

- Security missions.

- Reconnoiter and establish OPs and LPs.

Limitations

The scout platoon conducts reconnaissance operations as part of a larger combined arms force. The scout platoon depends on its parent unit for combat support and combat service support to augment and sustain its operations. Remember, the scout platoon's design and training is to reconnoiter platoon size areas only.

In addition, a full-strength platoon:

- Reconnoiters only a single route during route reconnaissance; METT-T is the determining factor.

- Reconnoiters a zone 3 to 5 kilometers wide; METT-T may increase or decrease the zone.

- During screening missions, is extremely limited in its ability to destroy or repel enemy reconnaissance units.

NOTE: Airborne and light infantry scouts are not mounted; they reconnoiter areas out 500 to 1,000 meters during most missions.

- During CR operations, can only acquire and maintain visual contact with the enemy; can kill or repel enemy reconnaissance elements only if augmented or task organized with infantry, armor, or AT assets.

- Operates six OPs for limited periods (under 12 hours), or three OPs for extended periods (over 12 hours). The light scout platoon usually operates three OPs.

- Is restricted in the distance it can operate from the main body, due to communications range and range of supporting elements.

- Cannot operate continuously on all battalion nets (such as battalion command, operations and intelligence, rear operations, and mortar) while operating on the platoon net. The platoon leader can monitor only two nets at the same time.

- Uses the HMMWV with only a 30-inch fording capability; its reconnaissance, surveillance, target acquisition, and night observation equipment includes the AN/UAS-11, Night Vision Sight; An/PVS-4, Night Vision Sight, Individual Weapon; AN/PVS-5, Night Vision Goggles; and binoculars.

Patrols (Maneuver Elements)

There are two basic categories of patrols: reconnaissance and combat. A patrol is a detachment sent to conduct reconnaissance, combat, or both. It consists of at least two people who may be accompanied by specially trained personnel or augmented with equipment essential to the mission. All maneuver elements conduct patrols during combat operations to provide reconnaissance, CR, security, and small-scale combat operations.

Reconnaissance patrols collect information and confirm or disprove the accuracy of previously gained information. The three types of reconnaissance patrols are route, area, and zone.

Combat patrols provide security and harass, destroy, or capture enemy personnel, equipment, and installations. The three types of combat patrols are raid, ambush, and security.

Capabilities

Patrols can conduct missions mounted or dismounted in various terrain and weather conditions. Patrols can also conduct missions 10 to 15 kilometers beyond the FEBA. Patrols sometimes pass through the scout platoon to conduct missions. Indirect fire should support patrols at all times. The distance for patrol missions varies with the type of patrol and METT-T. The company must always coordinate with the battalion before the patrols departure to eliminate redundancy and gaps.

Limitations

Patrols have many of the same limitations as the scout platoon. Patrols normally do not provide surveillance for extended periods. Patrols can reconnoiter an area, establish OPs/LPs for a limited period, and then leave.

Observation Posts/ Listening Posts (Maneuver Elements)

Units establish OPs/LPs to provide security, surveillance of NAI, and early warning of enemy activities. They are in use extensively during limited visibility. Proper emplacement includes concealment and unit support by fire.

Patrols differ in training and logistic support from scout platoons and normally do not establish OPs/LPs for extended periods. If you use patrols to conduct surveillance for extended periods, you are mismanaging your R&S assets. OPs/LPs are tasked to provide surveillance for extended periods as long as they meet the requirements stated above.

Capabilities

Units can employ practically an unlimited number of OPs/LPs. They can provide 24-hour coverage if they have the proper day and night observation devices, GSRs, or sensors. They can remain undetected due to lack of movement. Units can use OPs/LPs all over the battlefield as long as they are provided with fire support.

Limitations

OPs/LPs cannot operate for 24 hours if they do not have the proper equipment. A security element must be near the OP/LP to provide support and security in a timely manner.

Soldiers

During combat, soldiers are scattered all over the battlefield; thus, they can provide a large quantity of real-time information. You must get involved in the training to increase the timeliness and accuracy of information reported. All soldiers, from private to general officer, must know how to properly send information up the chain.

Capabilities

Soldiers can determine the types and numbers of enemy approaching.

Limitations

Soldiers do not always have the right equipment to send information quickly.

Equipment

Night observation devices (NODS) are either active or passive equipment designed to permit observation during darkness. Active equipment transmits infrared or white light to illuminate the target. Passive devices use either ambient light (from the stars, moon, or other low-intensity illumination) or operate by detecting the differences in heat (infrared energy) radiated by different objects. Heavy rain, snow, fog, or smoke degrade the effectiveness of these devices. You should use NODS on night patrols and OPs/LPs. Figure 3-1 shows observation equipment associated with the maneuver battalion.

DEVICE	CAPABILITIES	CHARACTERISTICS	ADVANTAGES AND DISADVANTAGES
AN/PVS-2 NV Individual Weapon	200 to 400 m	Weight 6 lbs 4 x magnification	Not detectable.
AN/TVS-2 NV Sight, Crew-served Weapons	1,000 m starlight, 1,200 m moonlight	Weight 15 lbs	Not detectable.
AN/TVS-4 NV	1,200 to 2,000 m	Weight 3.6 lbs 7 x magnification	Not detachable.
AN/PVS-4 NV Sight, Individual Weapons	400 m starlight, 600 m moonlight	Weight 3.6 lbs	Not detectable.
AN/TVS-5 Sight, Crew-served Weapons	1,000 m starlight, 1,200 m moonlight	Weight 7 lbs	Not detectable.
AN/PVS-5 NVG	150 m	Weight 1.9 lbs	Not detectable. Eye fatigue after 3 to 5 hours.
An/TAS-5 Thermal Dragon Sight light	1,000 m (+)	Weight 22 lbs	Penetrates all conditions of limited visibility and foliage. Not detectable. Short battery and coolant bottle life.
AN/UAS-12 Thermal TOW Sight	3,000 m (+)	Weight 18.7 lbs	Same as AN/TAS-5.

Figure 3-1. Observation equipment associated with the maneuver battalion.

DEVICE	CAPABILITIES	CHARACTERISTICS	ADVANTAGES AND DISADVANTAGES
AN/UAS-11 Thermal NOD	3,000 m (+)	Weight 58.4 lbs w/tripod	Same as AN/TAS-5.
Binoculars	Intensifies natural light.	7 x 50 power or 6 x 30 power	Not detectable. Requires some type of visible light.
AN/PAQ-4 Infrared Aiming Light	150 m (limited by PVS-5)	Weight .5 lbs Used with AN/PVS-5 and mounts on M16.	Permits aimed fire during darkness. Detectable.
AN/PAS-7 Hand-Held Thermal Viewer	Detects personnel at 400 m, at 1 km.	Weight 9.5 lbs	Penetrates all conditions of limited visibility and light foliage. Not detectable.

Figure 3-1. Observation equipment associated with the maneuver battalion (continued).

Assets and Personnel Normally Supporting the Maneuver Battalion

Assets and personnel that normally support the maneuver battalion include GSR, REMBASS, field artillery, engineer platoon, air defense artillery platoon, Army aviation, and tactical Air Force.

Ground Surveillance Radar

GSR provides the tactical commander with timely combat information and target acquisition data. The primary capability of GSR is to search, detect, and locate moving objects during limited visibility. GSR is capable of accurately locating targets for rapid engagement. It provides early warning of enemy movement and assists friendly forces in movement control.

Tasks

GSR is used to:

- Detect enemy movement during limited visibility.

- Monitor NAI.

- Monitor barriers and obstacles to detect enemy breaching.

- Monitor flanks.

- Extend the capabilities of patrols and OPs/LPs.

- Vector patrols.

- During daylight, detect enemy obscured by haze, smoke, or fog.

- Monitor possible drop zones or landing zones.

Capabilities
GSRs can:

- Penetrate smoke, haze, fog, light rain and snow, and light foliage.

- Operate in complete darkness.

- Detect moving personnel and equipment.

- Be moved around on the battlefield.

- Provide adjustment of indirect fire.

Limitations
GSR limitations are:

- Emits active radar waves which are subject to enemy detection and electronic countermeasures (ECM).

- Performance is degraded by heavy rain or snow and dense foliage.

- Line of sight (LOS) operation only.

- Limited mobility of the AN/PPS-5.

- Limited range of the AN/PPS-15.

Characteristics
GSR should be used with NODS as complementary surveillance devices, since each device can be used to overcome the limitations of the other. Figure 3-2 shows GSR characteristics. GSRs are organic to the MI battalion, intelligence and surveillance (I&S) company. The MI Battalion provides GSRs in direct support (DS) of brigade operations.

	AN/PPS-5	AN/PPS-15
RANGE: Personnel Vehicles	6,000 m 10,000 m	1,500 m 3,000 m
ACCURACY: Range Azimuth	±20 m ±10 mils	±20 m ±10 mils
* SECTOR SCAN:	Automatic - 553, 1,067, 1,600, and 1,955 mils (selectable)	Automatic 800 or 1,600 mils
INDICATORS:	Audio and visual (A- and B- scope)	Audio and visual (digital readout)
REMOTE CAPABILITY:	15.24 m	9.144 m

★ Both radars can be manually rotated to any azimuth and manual scanning can be performed.

Figure 3-2. GSR characterstics.

GSR teams that are DS to the brigade can be attached to maneuver battalion and company elements to support the commanders.

Radar Allocation

Radar is allocated as follows:

- Heavy Division:
  - Three squads of four teams each.
  - One PPS-5 per team equals 12 PPS-5s.
- Light Division:
  - Four squads of three teams each.
  - One PPS-15 per team equals 12 PPS-15s.
- Airborne division and air assault division:
  - Three squads of four teams each.
  - Three PPS-15s per squad equal nine PPS-15s (two-person team).
  - One PPS-5 per squad equals three PPS-5S (three-person team).

Site Selection Factors

General site selection should be made in close coordination with the GSR team leader whenever possible; specific site selection should always be left to the team leader. This takes advantage of the team leader's expertise and knowledge of the GSR. Site selection should provide:

- Protection by combat elements, as far forward as possible to provide the earliest warning.

- LOS between radar and target.

- Communication capability.

- Concealment and cover.

- Protection against ECM.

Remember, radars are an extreme electronic security risk. Both the main and side lobes emit sufficient energy for the enemy to detect and use radio ECM. GSRs, once detected, can give indicators to the enemy showing the size and disposition of friendly elements. GSRs can be destroyed or jammed. The following are common-sense OPSEC measures to be used with GSRs:

- Use terrain or vegetation to absorb or scatter side lobes.

- Place radar site so the target is between the radar and the hills or forests to limit the detection range.

- Schedule random operating periods of short duration.

GSRs can be used in tandem with two or more widely dispersed radars having the capability to illuminate the same target area, alternating operation times. The GSR can also be used with a night vision device that may not have the same range capability, but will provide some coverage when the radar is turned off.

REMBASS

REMBASS is organic to the airborne, air assault, and light division MI battalion, I&S company. REMBASS can remain under division, in general support (GS), or the division can provide it in DS to maneuver brigades, division support command headquarters, armored cavalry squadron, or maneuver battalion. REMBASS teams report directly to the G2 or S2 of the supported unit. The sensor monitoring set, which functions as the sensor output display, provides target identification and classification. In most cases, the sensor monitoring set is placed at the supported unit's TOC.

REMBASS teams hand deploy the sensors and repeaters; they also provide personnel to operate a monitoring device. REMBASS allocations are different for all divisions and are based on each division's particular mission. It is important to remember to include the REMBASS team leader in planning REMBASS missions.

Capabilities

REMBASS is an all-weather, day or night surveillance system, activated by magnetic, seismic-acoustic, or infrared changes from moving targets. REMBASS transmits target data by FM radio link to the monitors. With this data the operator can determine the target's:

- Direction of travel.
- Rate of speed.
- Length of column.
- Approximate number.
- Type (Personnel or wheeled or tracked vehicles).

REMBASS can operate in unusual climatic conditions and on varied terrain. REMBASS has transmission ranges of 15 kilometers (ground-to-ground), and 100 kilometers (ground-to-air). Because of the flexibility and wide range of REMBASS application, various sensor combinations can be selected to suit any given mission.

Limitations

Hand emplacement of sensors and repeaters in hostile areas increases employment response time. The sensor requires radio LOS to transmit data to the monitor. The equipment's weight and size limit the amount and distance personnel can hand carry for emplacement. REMBASS receivers are highly susceptible to electronic jamming; barrage jamming is the most effective. Operator proficiency greatly affects the results obtained.

Equipment

REMBASS teams normally use three different types of sensors: magnetic, seismic-acoustic, and infrared-passive. The sensors are arrayed in strings which complement one another. The sensors function automatically, transmitting information when movement, sound, or heat activates them.

Each sensor has detection and classification techniques suited to the physical disturbance (such as magnetic, seismic-acoustic, infrared-passive). Each sensor has a self-disabling and anti-tampering feature built into it.

Experience during Operation Desert Shield indicates an increased radius of detection for sensors emplaced in sand or sandy soil with a silica base, while sensors emplaced in loose rocky soil degrades sensor detection radius. Therefore, it is very important to check the detection radius of each sensor in the type soil of its intended employment and annotate the results on the Sensor Operator Data Record, if the situation permits.

Magnetic sensor. The magnetic sensor uses a passive magnetic technique to detect targets and determine the direction of movement (left to right and right to left). This sensor detects moving objects that are at least partially made of ferrous materials. The magnetic sensor will not classify targets. The magnetic sensor is most effectively used as a count indicator for vehicles.

Detection ranges of the magnetic sensor are:

- Armed personnel, 3 meters.

- Wheeled vehicles, 15 meters.

- Tracked vehicles, 25 meters.

Due to these detection ranges, REMBASS teams must use these sensors within proximity of the expected routes of travel. The weight of this sensor and battery is 3 kilograms.

Seismic-acoustic sensor. The seismic-acoustic sensor detects and classifies personnel and wheeled or tracked vehicles by analyzing target signature. It transmits a target classification report to the monitor. The weight of the sensor and its battery is 3 kilograms.

Detection ranges for the seismic-acoustic sensor are:

- Personnel, 50 meters.

- Wheeled vehicles, 250 meters.

- Tracked vehicles, 350 meters.

Infrared-passive sensor. This sensor detects and responds to a temperature change of 1.5 degrees Celsius within its field of view. It can determine the direction of motion relative to the sensor position. The infrared-passive sensor is most effectively used as a count indicator for personnel. The weight of the sensor and battery is 3 kilograms.

Detection ranges of the infrared-passive sensor are:

- Personnel, 3 to 20 meters.

- Vehicles, 3 to 50 meters.

Radio repeater. The radio repeater relays data transmissions between the sensors and the monitoring sites. The radio repeater intercepts the encoded radio message from either a REMBASS sensor or another like repeater.

Ranges of the repeater are:

- 15 kilometers ground-to-ground.

- 100 kilometers ground-to-air.

The repeater, like the sensors, has a self-disabling and anti-tampering feature built into it. The weight of the repeater and three batteries is 15 kilograms.

Additional equipment. Additional equipment for the sensor includes:

- A code programmer for programming a sensor or repeater to a desired operating channel.

- The antenna group for the REMBASS sensor monitoring set receives transmissions from extended ranges.

- Sensor monitoring sets for monitoring REMBASS radio-linked sensor and repeater transmissions.

- A portable radio frequency monitor to monitor sensors and repeaters. It is used primarily during emplacement of sensors to test operational status and radio LOS. It can also be used as a backup if the sensor monitoring set becomes inoperative.

Figure 3-3 shows site symbols. Adjacent brigades or battalions can monitor the same sensors if they exchange radio frequency information. This lateral monitoring increases the surveillance of units and promotes the exchange of intelligence. For additional information on REMBASS characteristics and employment techniques, refer to FM 34-10-1.

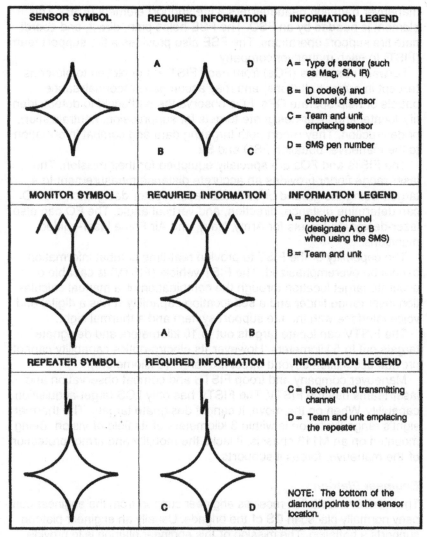

SENSOR SYMBOL	REQUIRED INFORMATION	INFORMATION LEGEND
	A D B C	A = Type of sensor (such as Mag, SA, IR) B = ID code(s) and channel of sensor C = Team and unit emplacing sensor D = SMS pen number
MONITOR SYMBOL	REQUIRED INFORMATION	INFORMATION LEGEND
	A B	A = Receiver channel (designate A or B when using the SMS) B = Team and unit
REPEATER SYMBOL	REQUIRED INFORMATION	INFORMATION LEGEND
	C D	C = Receiver and transmitting channel D = Team and unit emplacing the repeater NOTE: The bottom of the diamond points to the sensor location.

Figure 3-3. Site symbols.

Field Artillery

To properly integrate field artillery assets into the R&S and CR plans, you must understand the capabilities and limitations of this equipment.

A field artillery battalion is both a producer and a consumer of combat information. Field artillery battalions in DS of brigades provide each maneuver battalion headquarters a fire support element (FSE). This element is headed by an FSO. The FSE helps plan, direct, and coordinate fire support operations. The FSE also provides a fire support team (FIST) to each maneuver company.

Forward observers (FOs) from each FIST are deployed to platoons (except in armor battalions) and may accompany reconnaissance patrols or help operate OPs. FOs observe the battlefield to detect, identify, locate, and laser-designate targets for suppression, neutralization, or destruction. They report both targeting data and combat information to the maneuver battalion FSO and S2.

The FISTs and FOs are specially equipped for their mission. The laser range finder provides an accurate distance measurement to a target. Using the ground or vehicular laser locator designator, the FO can determine distance, direction, and vertical angle. The FO can also laser-designate targets for Army, Navy, and Air Force laser-guided munitions.

The capability of the FIST to provide real-time combat information cannot be overemphasized. The FIST vehicle (FISTV) is capable of accurate target location through the combination of a manual calculation laser range finder and a self-location capability. It has a digital and voice interface with the fire support system and a thermal sight.

The FISTV can locate targets out to 10 kilometers and designate targets out to 5 kilometers. However, its electro-optics capability cannot provide acquisition beyond direct fire range (5 kilometers).

Maneuver company and troop FISTS and combat observation and laser teams use the FISTV. The FISTV has only LOS target-acquisition capability. When on the move, it cannot designate targets. The thermal sight's range limitation is within 3 kilometers of its field of vision. Being mounted on an M113 chassis, it lacks the mobility and armor protection of the maneuver forces it supports.

Engineer Platoon

The maneuver battalion receives engineer support from the engineer company normally placed in DS of the brigade. Usually an engineer platoon supports a battalion. The mission of this engineer platoon is to provide mobility, countermobility, survivability, and general engineering support.

As the S2, you should learn to tap into this valuable resource for detailed information on natural and constructed terrain features. The S3 and the engineer support officer need to coordinate with each other to integrate engineer assets into the R&S and CR plans. The engineer support officer can provide key information about the terrain without your having to send out a reconnaissance patrol. In some cases it would prove valuable to have engineers go along with reconnaissance patrols. Engineers can provide expert terrain and obstacle analysis.

Air Defense Artillery Platoon

Short-range air defense elements normally support the maneuver battalion. These may include Stinger teams or sections and Vulcan squads. The air defense artillery (ADA) platoon or section leader functions as the battalion air defense officer. The ADA leader works closely with the battalion S2, S3-air, FSO, and air liaison officer (ALO) to plan and coordinate air defense support.

Specifically, the ADA leader would coordinate with you to pinpoint areas of enemy air and ground activity. The battalion air defense officer can tap into resources that look through the battlefield to determine areas of enemy air activity, thus revealing enemy ground activity.

The forward area alerting radar (FAAR) and target data display set provide air alert warning information to Vulcan squads and Stinger teams. This warning includes tentative identification, approximate range, and azimuth of approaching low-altitude aircraft out to 20 kilometers.

Army Aviation

Aviation units support maneuver brigade and battalion commanders. They provide a responsive, mobile, and extremely flexible means to find, fix, disrupt, and destroy enemy forces and their supporting command, control, and communications (C^3) facilities. Some aviation assets are capable of performing limited reconnaissance missions; however, most will collect information only as part of normal aviation missions.

You can find out from the Army aviation support officer information concerning enemy activity in areas where aviation assets fly missions. Helicopters can resupply, insert, or extract OPs/LPs or patrols. Combat aviation companies provide airlift support for troops and evacuate equipment, casualties, and enemy prisoners of war (EPWs).

The OH-58D is found in the attack helicopter battalion supporting maneuver brigades and battalions. The crew of the OH-58D consists of a pilot and an artillery fire support coordinator and observer. This aircraft performs two functions. First, when in support of maneuver battalions with FSEs, it is primarily a target acquisition and target attack system. Second, when in support of units where no FSE exists, the OH-58D crew performs a fire support planning and coordination function.

The OH-58D has many of the same capabilities as the FISTV. It has a thermal sight; a laser range finder and designator; a self-location capability; and a digital and voice interface with the fire support C^3 system. It can locate and designate targets out to 10 kilometers. Under less than ideal weather conditions it can only detect and recognize targets to within direct fire ranges.

This system provides:

- Digital interface with fire support C systems.

- Digital interface with Army aviation aircraft equipped with the airborne target handover system (such as the AH-64 Apache).

- Interface with Air Force assets so equipped.

These characteristics make the OH-58D a primary member of Joint Air Attack Team (JAAT) operations.

Tactical Air Force

JAAT is a combination of Army attack and scout helicopters and Air Force close air support (CAS). It normally operates in support of maneuver brigade or battalions. All staff officers participate in planning missions for Air Force support, especially the S2, S3, S3-air, FSO, and ALO. Coordinate through the ALO to receive real-time information from these Air Force assets.

The ALO also provides the means to forward immediate tactical air reconnaissance requests up the chain. Air reconnaissance reports, in-flight combat information reports, and air situation reports are all available through the ALO. The ALO weighs this information against information from the CR and the R&S plans. In this way the ALO can confirm or deny the accuracy of those plans.

Assets and Personnel Normally Supporting the Maneuver Brigade

At brigade level, the commander is fighting the close-in battle. You, as the brigade S2, must support the commander. To do this you must focus your collection effort at the 1st-echelon regiments and the 2d-echelon regiments capable of influencing your commander's battle (out to 30 kilometers). The brigade has limited assets available to conduct the collection effort. Here are some assets and personnel you can use to enhance your R&S and CR operations.

IEWSE

The IEWSE officer provides expertise on the capabilities, limitations, and employment of the intelligence and electronic warfare (IEW) equipment supporting the brigade. The IEWSE:

54

- Coordinates IEW support of the maneuver brigade.

- Is the link to the MI battalion for support.

- Communicates with the MI battalion (bn) to receive targeting and tasking information.

EPW Interrogators

Interrogators screen and interrogate EPW, detainees, and refugees. Their mission is to collect and report all information possible to satisfy the commander's PIR and IR. FM 34-80 contains the types of information interrogators can obtain and provide.

Counterintelligence

The counterintelligence (CI) support team can evaluate the vulnerability of friendly R&S assets to detection by threat R&S and target acquisition assets. CI members of support teams can identify and counter the specific enemy target acquisition means which pose a significant threat to brigade operations. These include:

- Human intelligence (HUMINT).

- Imagery intelligence (IMINT).

- SIGINT.

Based on enemy R&S activities, you could determine which operations security (OPSEC) and deception operations would work against the enemy, after coordination with CI personnel. FM 34-80 has detailed information on CI support.

GSR and Rembass

GSR and REMBASS can be kept under brigade control. See the above paragraph titled "Assets and Personnel Normally Supporting The Maneuver Battalion" for specific information.

Electronic Warfare Collection Systems

These assets operate near or within the brigade AO. They provide intelligence from intercepted enemy emitters. Ground-based systems include the following:

- AN/TSQ-138 (TRAILBLAZER) is a ground-based HF and VHF communications intercept and VHF DF system. It is found in heavy divisions only.

- AN/MSQ-103C, Heavy Divisions, AN/MSQ-103B, Airborne and Air Assault Divisions, (TEAMPACK) is a ground-based noncommunications intercept and line-of-bearing (LOB) system. It is found in all divisions except light.

- AN/TLQ-17A (TRAFFIC JAM) is a ground-based HF and VHF communications intercept and jamming system. It is found in all divisions except light.

- AN/TRQ-32 (TEAMMATE) is a ground-based HF, VHF, and UHF communications intercept and LOB system. It is found in all divisions.

- The AN/PRD-10\ll/12 is a ground-based radio DF system capable of being carried by one person. It is found in air assault, airborne, light divisions, and some heavy divisions.

- AN/ARQ-33A or AN/ALQ-151 (QUICKFIX) is an airborne communications intercept, jamming, and DF system. It is found in all divisions.

There are also numerous communications jamming resources that will be available to the commander in or near your brigade AO. FM 34-80 has detailed information about electronic warfare (EW) equipment.

Division Intelligence Officer
The G2 can provide many kinds of detailed intelligence. The G2 has various assets available to collect information and can pass this down to you as intelligence. FM 34-10 has specific information concerning these assets.

Field Artillery
Like the maneuver battalion, the maneuver brigade has an FSO to coordinate fire support. The FSO can communicate with numerous weapon-locating radars.

Fire Finder Radar
The target acquisition battery of division artillery (DIVARTY) has three AN/TPQ-36 mortar-locating radars and two AN/TPQ-37 artillery-locating radars.

The AN/TPQ-36 detects mortars and artillery out to 12 kilometers and detects rockets out to 24 kilometers.

The AN/TPQ-37 detects artillery and mortars out to 30 kilometers and rockets out to 50 kilometers.

Moving Target Locating Radars

The DIVARTY target acquisition battery has either one AN/TPS-25A or one AN/TPS-58B moving-target-locating radar. These battlefield surveillance radars are similar to the GSR. They can detect, locate, and distinguish wheeled and tracked vehicles and dismounted personnel.

The AN/TPS-25A detects moving vehicles out to 18 kilometers and personnel out to 12 kilometers.

The AN/TPS-58B detects moving vehicles out to 20 kilometers and personnel out to 10 kilometers.

Field Artillery Battalion Observation Posts

Survey parties and other trained personnel of the field artillery battalion operate the battalion OPs. These personnel survey designated target areas, and record high-burst and mean point-of-impact registrations. They send targeting data and combat information to the fire direction center or the FSO at the maneuver battalion or brigade.

Aerial Fire Support Officer

The DIVARTY support platoon of the heavy division's combat aviation brigade provides rotary wing aircraft for DIVARTY air observers. Their mission is to call for or adjust fires from the fire support assets. Aerial fire support officers:

- Cover areas masked from ground observers.

- Cover thinly resourced areas.

- Provide coverage while ground-based R&S and target acquisition assets displace.

- Reinforce surveillance of vulnerable areas.

- Report targeting data and combat information to the FSO at the maneuver battalion or brigade, DIVARTY TOC, or the fire direction center.

Engineer and Air Defense Artillery

Engineer and ADA support officers are located at the maneuver brigade. Types of information these personnel can provide is discussed in the above paragraph titled "Assets and Personnel Normally Supporting The Maneuver Battalion."

Air and Armored Cavalry Squadron

This squadron supports the division by conducting reconnaissance and security missions. There are four types of air and armored cavalry squadrons:

- The air and armored cavalry squadron of the heavy division consists of two ground cavalry troops (M3 equipped) and two air cavalry troops (OH-58s and attack helicopters).

- The air cavalry squadron of the air assault division consists of three air cavalry troops and one air assault troop.

- The air cavalry squadron of the airborne division consists of three air cavalry troops, one ground cavalry troop (tube-launched, optically tracked, wire guided [TOW] missile systems and scout HMMWVs), and one air assault troop.

- The air cavalry squadron of the light division consists of two air cavalry troops and one ground cavalry troop (TOW and scout HMMWVs).

Headquarters and headquarters troops and maintenance troops are not included in the above list.

Army Aviation

Attack Helicopter Battalions

These battalions are primarily trained to "kill" enemy tanks. They can also:

- Provide aerial escort and suppressive fires to support air assault operations.

- Destroy enemy C^3 and logistic assets.

- Conduct JAAT operations.

If these assets support your brigade, they can provide detailed information about enemy activity. The key to obtaining this information is to coordinate with the S3-air and the Army aviation support officer. Refer to the above paragraph titled "Assets and Personnel Normally Supporting The Maneuver Battalion" for additional information. Figure 3-4 shows an asset deployment matrix. This matrix may be used by brigade and battalion S2s to keep track of deployed assets.

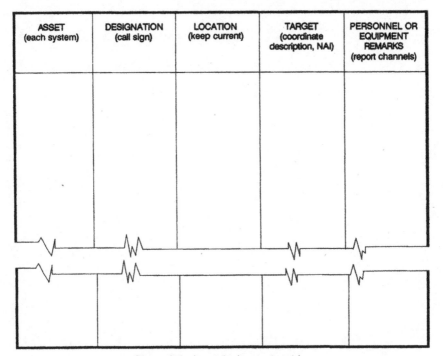

ASSET (each system)	DESIGNATION (call sign)	LOCATION (keep current)	TARGET (coordinate description, NAI)	PERSONNEL OR EQUIPMENT REMARKS (report channels)

Figure 3-4. Asset deployment matrix.

Military Police Platoon

The military police (MP) platoon supports the maneuver brigade during some missions. If you have an MP platoon supporting your unit, you should coordinate with the MP platoon leader for information. The MP platoon can usually coordinate with other MPs who are normally scattered all over the AO.

CHAPTER 4

Planning Effective Reconnaissance and Surveillance

This chapter presents the planning steps for effective R&S operations. The S2 is responsible for making recommendations in R&S operations. At brigade or battalion, you are the driving force in the R&S effort. (Refer to Chapter 1 for the collection management process.) These steps apply to both brigade and battalion levels.

The first step begins with receiving the unit's mission. You must understand the commander's intent in this particular mission. You have already completed most of the IPB process, but now you must produce some of the specific information pertaining to the mission.

Once you understand the mission, begin to analyze the requirements placed on you as the S2. The commander should tell you the key pieces of information needed before and during the mission. This key information, called PIR, is either stated by the commander or recommended by you for the commander's approval.

The PIR and IR provide the initial focus of the R&S effort. The R&S plan should answer the PIR and IR. At this point you should have a rough draft R&S plan, such as when and what areas to begin R&S operations. (This is part of the mission analysis phase of the planning process steps.)

You can now begin adding some detail to the R&S plan. Integrate any requirements from higher headquarters into the plan. You have to translate the initial PIR and IR into indicators on which R&S assets can collect. Figure 4-1 shows examples of the PIR, indicator, specific information requirements (SIR), and specific orders and requests (SOR) process. Additional examples of indicators are in FM 34-3, Appendix C. Now determine the SIR and SOR needed for the R&S plan. The SIR and SOR ensure assets are collecting specific information that answers the PIR and IR.

The event template is a product of IPB showing when and where the enemy could go. Compare the SIR to the event template; this comparison should indicate when and where to send friendly R&S assets. Those areas in which you expect enemy activity are labeled NAI.

ENEMY OFFENSE

PIR: "Will the enemy launch a regimental- or battalion-size attack within my sector in the next 12 hours?" - Commander

Indicator: Reconnaissance and destruction of our defensive obstacles.

NOTE: Usually the night preceding an attack, enemy patrols reconnoiter friendly obstacles to determine a plan for clearing lanes. The patrol destroys only those obstacles that will not disclose the direction of the main attack. - Threat Knowledge

SIR: "Is the enemy conducting reconnaissance and clearing obstacles along our FLOT?" - S2

SOR: "Report on status of friendly obstacles and the area around the obstacles. Report the location of all destroyed friendly obstacles or the indication of recent enemy activity around the obstacles (such as tracks, footprints, or crushed vegetation)." - S2

ENEMY DEFENSE

PIR: "Is the enemy going to defend or will they continue the attack?" - Commander

Indicator: Formation of AT strong points in-depth along logical AAs.

NOTE: The enemy forms AT strong points in-depth along logical AAs for armor. These are made up of motorized rifle, engineers, and AT gun and missile units strengthened by mines, ditches, and other obstacles. - Threat Knowledge

SIR: "Are the enemy placing their AT weapons forward as well as building obstacles?" - S2

SOR: "Reconnoiter NAI 4. Specifically look for and report enemy AT positions (AT 4 or 5 mounted on BMPs or BRDMs), BMPs, and obstacles." - S2

Figure 4-1. Example of the PIR, indicator, SIR, and SOR process.

Once you have a picture of the coverage required for the R&S effort, you should prioritize the SIR. Those SIR that, when answered, will provide the greatest amount of intelligence in the shortest amount of time should have a high priority. (This is part of the COA development phase of the planning process steps.)

By now you have set your SIR priorities, identified areas to send R&S assets, and know when to begin the R&S mission. For the next step, you must be familiar with the capabilities and limitations of all R&S assets at your particular level. Compare the SIR with available R&S assets. Close coordination between you and the other staff officers should help ensure the assets are properly deployed. Development of the R&S plan should involve all staff officers. Your concern is developing IR and guiding assets to the proper areas.

Staff Officer Responsibilities

Other staff officers have a role in this process. The following is a list of these officers and their responsibilities:

- The S3 makes sure the assets are available and can conduct the mission and the R&S plan supports the overall mission of the unit.

- The CI officer apprises you of the vulnerability of your R&S assets to enemy collection and target capabilities.

- The FSO coordinates indirect fires planned to support R&S assets and recommends establishing appropriate restrictive fire support coordination measures to provide for troop safety.

- The R&S asset commander is responsible for planning targets and indirect fires for that element.

- The engineer officer supports the R&S effort by collecting information on the terrain and obstacles.

- The IEWSE officer supports the R&S effort by guiding the MI battalion assets to assist in answering the PIR.

- The ADA officer plans air defense for the R&S assets and also provides information on enemy air activity.

- The NBC officer integrates NBC operations with R&S missions.

- The aviation officer provides air movement for R&S assets and also information on enemy activity while in flight.

- The ALO provides close air support for R&S missions as well as in-flight reports on enemy movement.

These staff officers are not cast aside upon completion of the R&S plan. They should be kept updated on the current R&S situation. These officers provide recommendations during R&S operations and ensure their assets are operating as instructed.

The DST is a tool used in the IPB process that brings the staff officers together to plan the mission. The DST also ensures involvement among the S2, the S3, and the FSO in planning R&S missions.

Planning

Once you know which R&S assets are available to conduct R&S operations, you have to decide how to satisfy the SIR. To collect the greatest amount of intelligence with the fewest assets, you must know how to plan missions using basic collection management strategy such as augmenting, task organizing, cueing, and redundancy.

Augmenting

Chapter 8 has an in-depth discussion of augmenting.

Task Organizing

To collect the most information with the fewest assets and in the quickest way, task organize assets. This increases their overall effectiveness in gathering information and surviving on the battlefield. The following is an example of task organizing.

A scout platoon's mission is to conduct a 10-kilometer-wide by 10-kilometer-deep zone reconnaissance before a movement to contact. The scout platoon must accomplish this mission in one hour. You have determined the platoon needs augmentation to cover this much area in the time allowed. After you coordinate with the S3, the S3 attaches two mechanized infantry squads to the scout platoon. These two squads are given the mission to provide security and mark infiltration routes.

In this example the scout platoon is able to concentrate on reconnoitering the terrain and locating enemy positions while the two mechanized squads provide security for the scout platoon and mark infiltration routes. If you had expected heavy enemy obstacles, the S3 could have attached an engineer section to mark, breach, or provide obstacle assessment while the scouts and infantry did their mission.

You should consider all the assets listed in Chapter 3 for augmentation or task organization roles. See Chapter 8 for further discussion of task organization.

Cueing

Another collection strategy of R&S missions is cueing. Cueing is using limited assets to identify or verify enemy activity or using one asset to tip off or alert another asset. Use the event template to pinpoint the times and areas to collect on the enemy.

Instead of the R&S assets trying to cover large areas for extended periods of time, the assets are active only when cued. The cueing can be the time you expect the enemy to be at a specific NAI, or the reaction to information reported by another asset. An example of cueing follows:

You have identified three NAI needing surveillance, while using only one asset. For this example the only asset available to cover the three NAI is an OP. Due to the distance between the NAI, the OP cannot cover all three NAI at the same time. You determine a location central to all three NAI. From this location the OP can cover only one NAI with surveillance.

An aircraft reports enemy vehicles near one of the unsurveilled NAI. You inform the OP of the activity, thus cueing it, and the asset moves toward the NAI to verify the report. You may use any of the assets listed in Chapter 3 as cues for other assets.

Redundancy

Another collection strategy for R&S operations is redundancy. As the S2, your primary effort is to provide R&S coverage for as many NAI as possible. Based on the priority of the SIR and the number of NAI, you have to decide which areas you want more than one asset to cover. With more than one asset covering the same NAI, a backup system is available in case one asset breaks down. Redundancy guarantees continuous area coverage. An example of redundancy follows:

You have a GSR covering an NAI during limited visibility. Just in case the GSR breaks down, you have assigned two OPs/LPs with NODS to cover the NAI. The OPs also provide NAI coverage during daylight while the GSR crew rests. If the GSR breaks down, the OPs have NODS to pick up the responsibility of surveilling the NAI. The NODS can also specifically identify the moving intruders detected by the GSR.

Remember to include in the R&S planning efforts coordination with the CI team supporting your unit. The routes used by your scout platoon and the positions operated by your assets will be potential NAI to enemy collection assets. Whatever OPSEC and deception measures you incorporate into the R&S plan, they should be based on CI evaluation of the vulnerability of your R&S assets to enemy collection and target acquisition capabilities. As you expect to see the enemy at certain times and places on the battlefield, so the enemy will expect to see you.

Working with the S3, you are now ready to begin matching assets with missions. If the commander and the S3 approve the R&S plan, then give warning orders to the assets. The warning orders allow the assets enough time to conduct troop-leading procedures.

Once you issue the warning orders and refined R&S plan, prepare your portion of the mission briefing. The purpose of this briefing is to inform the collection assets of their missions and to provide them with as much information as possible about it (such as IPB products). Use all available information to provide as clear a picture as possible of what you expect of them on the battlefield and what they can expect to encounter.

CHAPTER 5

Methods of Tasking Reconnaissance and Surveillance Assets

The S3 is responsible for and has tasking authority over maneuver elements. The S2 makes tasking recommendations to the commander or S3. The S2 is the primary user of the scouts and, many times, the S2 actually tasks them. In some units the S2 has tasking authority over R&S assets after the commander and the S3 approve the R&S plan. Every unit has its own SOP regarding R&S responsibilities.

Normally, you can consider the commander's approval of the R&S plan as granting tasking authority. Many times the S3 does not have time to prepare and publish separate R&S missions, so the S2 does it. Once the S2, after coordinating closely with the S3, has completed writing the formal plan, it is sent to the commander for approval. Once the commander signs the fragmentary order (FRAGO) or warning order, the taskings within it become missions ordered by the commander, regardless of who actually wrote it.

Another way to publish the missions and taskings of the R&S plan is to print it in the subordinate unit instructions within the unit OPORD. While not as timely as FRAGOs or warning orders, unit instructions still let everyone know what you expect of them during the R&S operations.

Figure 5-1 shows a sample R&S tasking with subordinate unit instructions. Maneuver battalions can also use this method when it is tailored to their level.

Probably the quickest way to issue R&S orders is to have the S3 issue them when issuing the warning order to the unit. This method ensures the tasked assets know it is a formal tasking coming from the commander. Both the S2 and S3 plan the R&S operation; however, the S3 continues to have the responsibility of actually tasking assets.

A formal method of tasking assets for R&S operations is an intelligence annex to the OPORD. The intelligence annex is a formal intelligence tasking document accompanying an OPORD or an operations plan (OPLAN). Paragraph 2, Priority Intelligence Requirements, and paragraph 3, Intelligence Acquisition Tasks, inform all assets what the focus of the R&S plan is and what mission each asset is to conduct. The S2 is responsible for the intelligence annex. Again, before distribution, the S2 coordinates the intelligence annex with the commander and S3 for their approval.

(Classification)

*   *   *   *   *   *   *

OPORD 19 3d Bde - 24th Armd Div...

3. Subordinate Unit Instructions

    b. TF 1-6:

        (1) Establish contact point 1 and assist passage of covering force elements through the FEBA.

        (2) Defend in sector from NT888777 to NT666555 NLT 181500U JAN94.

        (3) Establish surveillance sites along Route TOM and HARRY to provide early warning of enemy flank attack NLT 181500U JAN94.

        (4) Send CR patrols along the FIBEL river and the DICKEL ridge line NLT 181600U JAN94 to confirm or deny the presence of enemy reconnaissance patrols.

    c. TF 1-68:

        (1) Establish contact point 2 and assist passage of covering force elements through the FEBA.

        (2) Defend in sector from NS666444 to NS333222 NLT 181500U JAN94.

        (3) Establish screen along the east side of the NEILSON AQUADUCT NLT 181500U JAN94. Establish surveillance positions near the intersection of AUBOBAHN 67 and 4 to provide early warning of enemy reinforcements.

        (4) Send a CR patrol in the forests southeast of KIBBLESNBITS capable of destroying a reinforced BMP Company NLT 181700U JAN94 to destroy enemy reconnaissance units.

(Classification)

Figure 5-1. Sample R&S tasking with subordinate unit instructions.

Figure 5-2 is a sample R&S tasking. Maneuver battalions can also use this method, but would tailor it to their level.

```
                        (Classification)

    *         *         *         *         *         *         *

ANNEX B (Intelligence) to OPORD 19...

REFERENCES:  Map sheets

TIME ZONE:  U

3.  Intelligence acquisition tasks.

    a.  TF 1-6:

        (1)  Establish surveillance sites along routes TOM and
HARRY to provide early warning of enemy flank attack NLT 181500U
JAN94.

        (2)  Send CR patrols along the FIBEL river and the
DICKEL ridge line NLT 181600U JAN94 to confirm or deny the
presence of enemy recon elements.

    b.  TF 1-68:

        (1)  Establish screen along the east side of the NEILSON
AQUADUCT NLT 181500U JAN94.  Establish surveillance positions
near the intersection of AUTOBAHN 67 and 4 to provide early
warning of enemy reinforcements.

        (2)  Send a CR patrol in the forests southeast of
KIBBLESNBITS capable of destroying a reinforced BMP Company
NLT 181700U JAN94.




                        (Classification)
```

Figure 5-2. Sample R&S tasking in the intelligence annex to the OPORD.

The R&S tasking matrix is another method of tasking R&S assets. (See Appendix A.) Distribution can be:

- Directly to the tasked asset.

- Attached the subordinate unit instructions of the OPORD.

- Attached the intelligence annex.

Once tasked, the assets must understand their mission. During the mission briefing for the assets, use as many of the IPB products as possible. Each product serves a particular purpose:

- The modified combined obstacles overlay (MCOO) reveals terrain constraints.

- Photographs show terrain features.

- Enemy situation templates provide a picture of the enemy's location and probable COA.

- Event templates indicate where and when the enemy can maneuver.

- NAI on the event template show target locations.

- The DST provides a picture of the overall friendly scheme of maneuver and warns the R&S assets of any friendly fires in their vicinity.

Once the assets have an understanding of the enemy and terrain, they can receive detailed mission instructions. It does not matter if you or the S3 give this briefing, just as long as the assets understand their mission. If possible, the S2, S3, and FSO should be involved in this briefing.

Each asset should understand what the commander wants it to accomplish. Give the assets the big picture, then direct them to their specific roles and how they are to support the overall mission.

You can see the extensive time required to plan R&S operations. In most units there is not enough time to go into a lot of mission-specific detailed planning before the assets are deployed. R&S operations must begin as soon as possible after the unit receives the warning order or OPORD. If this is the case, assets can be sent out after you and the S3 have developed the rough draft R&S plan. Later, as you and the other staff officers refine the R&S plan, you can adjust the assets and their instructions.

CHAPTER 6

The Reconnaissance and Surveillance Overlay

The R&S overlay is the R&S plan in graphic form. The purpose of the R&S overlay is to show the assets and the key staff officers exactly where the R&S assets are operating. You will extract most of the overlay's graphics and symbols from FM 101-5-1. Additionally, due to the various R&S operational techniques, you will need to construct some "homemade" graphics and explain them in the legend.

There are two parts to the R&S overlay. The first part is the graphic display of deployed or planned deployment of R&S assets. The second part is the marginal data consisting of the legend, administrative data, specific instructions to each asset, and the distribution list.

The marginal information found on the overlay consists of the standard wording found on all overlays. The administrative data is comprised of the following:

- Classification.

- Overlay title.

- Registration marks.

- Map sheet name.

- Map sheet scale.

- Map sheet number.

- Map sheet series.

- The "prepared by" line.

Another portion of the administrative data is the legend. The legend contains any nonstandard FM 101-5-1 graphics used. It also contains detailed written instructions to each R&S asset. These detailed instructions should focus on:

- The required operational times. You should give each asset both a start and a finish time for each mission, as applicable.

- The target. To answer the PIR, you need to look for specific indicators. Each asset should be told exactly what to look for (such as type units, equipment, and specific activity). Never give broad-based generic missions to "go out and look for and report on anything that moves." Specific guidance will promote specific answers.

- Coordinating instructions. All assets will, at one time or another, move through or near another unit's AO. To keep units from shooting friendly R&S assets, assets and units must coordinate with each other. It is also important that R&S assets coordinate among themselves.

- Reporting requirements. All assets should know when, how often, and what format to use when reporting. You should provide frequencies, alternate frequencies, and reaction during jamming. You must also provide the NLT time for specific information to be reported.

Initially, the locations for assets are areas in which you recommend they deploy. After the assets have gone to these areas (NAI), they report to you or the S3 the actual locations in which they can conduct their missions. You or the S3 updates the graphics to show actual locations.
Control measures are as follows:

- Friendly boundaries, R&S limit of responsibility, NAI, start points (SP), release points (RP), and checkpoints.

- Graphics depicting route, area, and zone reconnaissance.

- Primary, alternate, and supplementary positions.

- Sectors of scan for sensors.

All of these control measures, except R&S limit of responsibility, are found in FM 101-5-1. The R&S limit of responsibility comes down from higher headquarters along with other R&S guidance.

This limit is a control measure that informs subordinate units of the limits of their R&S operations. It can be represented by a dashed line (- - - -).

Remember, it is important to include detailed instructions for each R&S asset on the overlay. This method is known as the overlay method for distributing written instructions. Another method is known as the matrix method. The R&S tasking matrix is the matrix used for this method. (See Appendix A.) Figure 6-1 shows an R&S tasking matrix.

Figure 6-2 is an example of the R&S plan graphically portrayed on an overlay with detailed instructions to each asset written on the bottom of the overlay. Attach the R&S tasking matrix to the bottom of the R&S overlay. The matrix provides the following information:

■ The first column shows the priority of each mission. This number should correspond with the PIR number.

■ The next column provides the asset with the NAI number and grid coordinate.

■ The start/stop column informs the asset the times for this mission.

■ The SIR column explains to the assets exactly what they are looking for (target).

■ The next set of columns lists the actual assets tasked to conduct each mission. An "X" placed under each asset identifies the tasking.

■ The coordination column tells the assets which units to coordinate with for this mission.

■ The last column provides the assets with reporting requirements.

We have discussed two ways to distribute instructions; however, the method is not important. What is important is for assets to receive clear, specific instructions.

Disseminating the R&S plan to all the assets can be a problem. When the R&S plan reaches the dissemination phase, the assets are usually scattered great distances over the battlefield. In some instances the R&S plan is disseminated by courier. To ensure all assets receive their copy, write each asset's title directly on the distribution list, plan, or

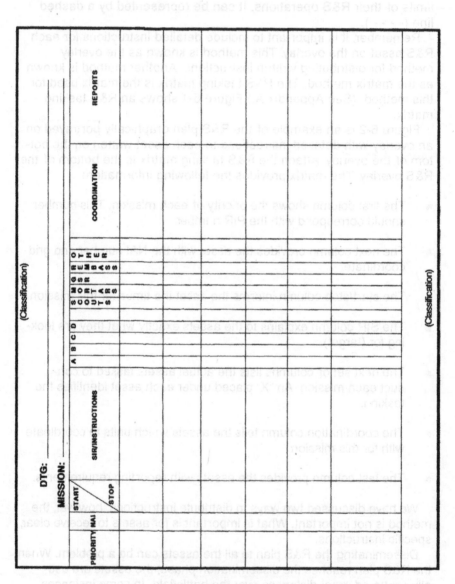

Figure 6-1. R&S tasking matrix.

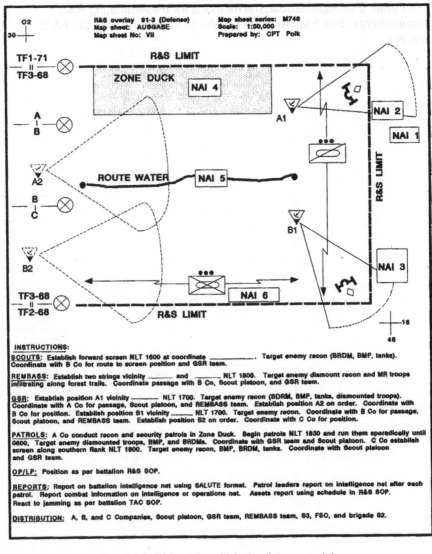

Figure 6-2. R&S overlay with instructions on acetate.

overlay. Exchanging brigade or battalion R&S plans with adjacent units ensures proper coordination, minimizes the risk of shooting friendly soldiers, and cuts out unnecessary redundancy.

Chapter 5 provides additional methods for disseminating R&S requirements. See FM 34-80, Appendix E, for another sample R&S overlay.

CHAPTER 7

Monitoring the Reconnaissance and Surveillance Effort

During the battle, your commander will probably have several PIR that need to be answered. As the S2, you will weigh those PIR in some way to organize your collection effort. Remember the first principle of R&S: tell commanders what they need to know in time for them to act.

This principle implies that, as the S2, you must monitor your collection effort at all times so you can make sure you answer your commander's PIR.

If your commander develops new PIR during the battle, you may have to modify your R&S plan to address the new PIR. Suppose a PIR becomes obsolete. For example, let's say your commander was very concerned about enemy reconnaissance locations. This PIR would be valid as your unit prepared to cross the LD/LC. However, once your unit consolidates on its objective, this particular PIR would be less important. The point is, you should constantly monitor the status of your R&S effort so you will know when to update PIR or to modify your R&S plan.

Tracking Targets and Assets

There are other reasons you must monitor your R&S or collection plan. Remember the term "high payoff target," or HPT? These are specific enemy weapon systems or specific enemy units that are identified which must be destroyed, degraded, or suppressed for your unit to succeed in its mission. Many times, locating an HPT may be one of your commander's PIR. Other times, it might be an IR. In either case, your R&S plan must account for HPTs. During the execution of your R&S plan, you must be able to identify HPTs and quickly forward their location to the S3 and FSO for action. This is especially critical for CR operations.

Another reason for monitoring your R&S operation is to keep track of your asset status and location. You will need to know which of your assets are still mission capable and which are inoperative, so you will not waste time retasking inoperable assets. Obviously, if you need to retask assets from one location to another, you need to know where those assets are.

One technique to keep track of your assets is to have them report in at predetermined intervals based on METT-T, criticality of the area covered by the asset, or communications available. You can even show this graphically by using TPLs for moving assets.

For example, let us say you have given your scouts the mission of route reconnaissance. On your event template, you have developed a series of TPLs depicting 15-minute increments. As your scouts cross a TPL, they report in to you. In this way, you can easily monitor where your scouts are on the battlefield. (Instead of TPLs, you can use existing friendly control graphics as well.)

If you lose contact with your scouts, you at least have an approximate idea of where they last were. When you use TPLs, try to have recognizable features represent them. Figure 7-1 is an example of this technique.

Evaluating How Your Assets Report

You should monitor your R&S plan to evaluate how well your assets are reporting information back to you. If your assets are not reporting quickly enough, accurately enough, or reporting the wrong information, you will need to make corrections.

At the brigade and battalion levels, many times you will find your assets may not always provide you with timely or complete information. There are many reasons for this. Most of the time it is difficult to discern what is happening on the battlefield. The company commander or platoon leader is preoccupied with fighting and winning the battle.

Nevertheless, do not accept incomplete information! If a spot report lacks the type of vehicle, number of vehicles, or direction of movement, get back on the radio and ask for it. If your scouts send back a report that does not make sense to you, ask for clarification. If you have not heard from your ground surveillance radars for an unusually long time, call them and ask for a situation report.

You should enforce negative situation reports at predetermined intervals. Too many times in the past, S2s thought no news was good news. They were content to sit in their vehicles in silence. Be aggressive! Remember, you are trying to answer your commander's questions. You cannot, and commanders cannot do their jobs, unless incoming information is timely, accurate, and complete. Nevertheless, you must be realistic.

There is much confusion in battle, and some information will not be attainable. You cannot tie up the radio nets trying to get "perfect" reports. Some information you will have to live without.

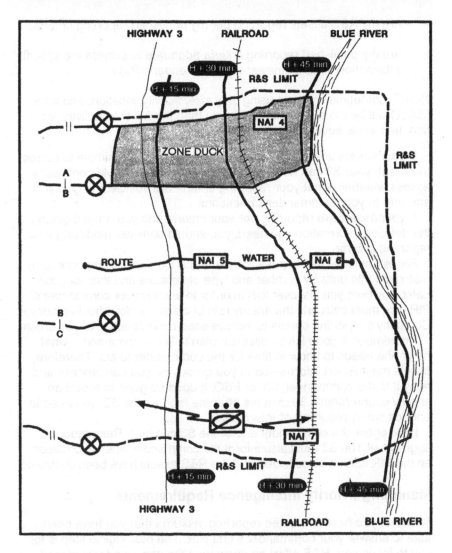

Figure 7-1. Time phase line control.

The reporting criteria you have specified in your intelligence annex or on your R&S overlay will tell your assets how and when they are to report. As you monitor your R&S operation, you should evaluate two things:

- Are my R&S assets reporting per my published reporting criteria?

- Are my published reporting criteria adequate to provide the specific information used to answer my commander's PIR?

NOTE: Sometimes reports using only size, activity, location, and time (SALT) will be more feasible and timely than the full size, activity, location, unit, time, equipment (SALUTE) report.

If your assets are not reporting per your criteria, it is simple to correct the asset; your S3 can help with this. However, you should continually assess whether or not your reporting criteria are sufficient to give you answers to your commander's questions.

If your assets are reporting per your criteria and you are not getting the detailed information you need, you should consider modifying your reporting criteria.

For example, let us suppose you wanted A Company to report enemy motorized rifle units by number and type of vehicles and their location. Later, however, you discover that in order to answer your commander's PIR, you must calculate the enemy rate of advance. You should modify A Company's reporting criteria to include speed and direction of movement.

Remember, a good R&S collection plan tells the commander what he or she needs to know in time for the commander to act. Therefore, assets must report information to you quickly so you can process and relay it to the commander, S3, or FSO. It does no good to report an enemy counterattack 30 minutes after the fact. As the S2, you need to enforce timely reporting of information.

Here again, the commander and/or the S3 can help. Remember, be aggressive! The S2 must also inform the commander when information on the PIR cannot be collected or if the R&S assets have been destroyed.

Managing Priority Intelligence Requirements

Now that you have evaluated reporting, assume that you have been able to answer your commander's first PIR. The next logical step is for you to focus your R&S effort on answering the commander's second highest PIR, then the third, fourth, and so on. Realistically, your R&S plan will probably address more than one PIR simultaneously. The point for you to remember is that R&S does not stop. Once you have satisfied a requirement, shift your attention to the next highest priority.

Many times you may have answered a PIR out of sequence. For example, you may be able to answer PIR 2 and 3 although you still have not been able to collect enough information to answer PIR 1. Or you may find the battlefield situation has changed so drastically your PIR 1 is no longer a valid concern.

These cases prove you must continually reevaluate the Priority of your commander's PIR. If you have answered PIR 2 and 3, does PIR 4 become your second priority? If PIR 1 is no longer a valid concern, does PIR 2 become your top priority? You must support your commander. Knowing and understanding your commander's intent will help you reevaluate priorities and anticipate possible changes, as will a solid relationship with your commander and S3.

One useful technique that will aid you in managing PIR priorities is to "time phase" your commander's PIR based on how you anticipate events on the battlefield. Essentially, you tie each PIR to a phase in the battle through use of the DST.

Normally, each PIR has a time relative to a point in the battle when answering it will be important, and another time when the PIR will no longer be a valid concern. For example, let us suppose your unit's mission is to attack. Initially, the most important thing your commander might need to know is the location of enemy reconnaissance and security zone units.

However, after a certain point in the attack (after you have penetrated the security zone), this question becomes meaningless. Now, the most important thing might be to locate the enemy's main defensive area. Once you have consolidated on the objective, the most important thing might be locating any possible enemy counterattack. Therefore, before the attack, your commander's PIR might look like this:

- PIR 1: What are the locations of the 34th motorized rifle regiment (MRR) reconnaissance and platoon strong points in the security zone?

- PIR 2: What are the locations of the 34th MRR's MRC and AT positions within the main defensive area?

- PIR 3: What is the location of the 4th Tank Battalion (TB) (-) of the 34th MRR?

Once you have reached your intermediate objective, you might change your commander's PIR priorities to look like this:

- PIR 1: What are the locations of the 34th MRR's MRC and AT positions within the main defensive area?

- PIR 2: What is the location of the 4th TB (-) of the 34th MRR?

- PIR 3: What are the locations of the 34th MRR's reconnaissance and platoon strong points in the security zone?

In fact, you might delete PIR 3 altogether. Later, as you consolidate on your subsequent objective, you may reprioritize like this:

- PIR 1: What is the location of the 4th TB (-) of the 34th MRR?

- PIR 2: What are the locations of the 34th MRR's MRC and AT positions within the main defensive area?

Since PIR relate to events on the battlefield, you can anticipate them by war gaming; and enter these changes onto the intelligence BOS of your unit's DST. Figure 7-2 shows the process of time phasing PIR.

Modifying the Reconnaissance and Surveillance Plan

Whether modifying reporting requirements because of new reporting criteria or because of new or modified PIR, you must be ready to change your R&S plan to fit the commander's needs. Basically, you will have to decide:

- Where you want your R&S assets to shift their attention.

- Where you want those assets to actually move.

- What you want your assets to look for.

- How you want your assets to report.

Here is where doing your homework (IPB) ahead of time comes in handy. If most or all of your IPB products were prepared ahead of time, all you need do is review and update those products as necessary.

Looking at your updated situation templates and event templates will give you a good idea of where to shift your R&S focus, and what you should expect to see. Your updated terrain and weather products will tell you where to place your assets.

However, if you have not been able to update or produce situation and event templates, or you have advanced past your AI, you still need to mentally envision what you think the enemy will look like on the terrain, applying the effects of weather. Your mental picture will help you quickly come up with NAI and TPLs.

The next step is to retask your R&S assets. Remember, when you shift your R&S assets, their vulnerability to enemy collection and target acquisition capabilities may change.

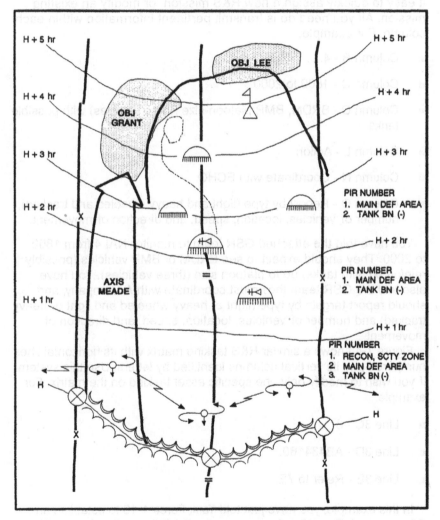

Figure 7-2. Time phasing PIR.

Tasking Assets

Chapter 5 described many ways of tasking assets, including using a matrix format. A matrix is easy to use and can be quickly modified. Figure 7-3 is an example of a modified matrix.

Each column has a letter designator. For example, the Priority column is "A," the NAI column is "B," and so on. The lettering makes it easy to quickly assign a new R&S mission, or modify an existing mission. All you need do is transmit pertinent information within each column. For example:

- Column B - 4.

- Column C - 1800 to 2000.

- Column D - BRDM, BMP, platoon-size (three vehicles) with possible tanks.

- Column L - Action.

- Column N - Coordinate with ECHO.

- Column O - Report by type (light and heavy wheeled and tracked), number of vehicles, location, speed, and direction of movement.

You have told the attached GSR team to monitor NAI 4 from 1800 to 2000. They should expect to see BRDM or BMP vehicles (possibly reinforced with tanks) up to platoon size (three vehicles). You have also told the GSR team they must coordinate with A Company, and should report targets by type (light or heavy wheeled and light or heavy tracked) and number of vehicles, location, speed, and direction of movement.

Figure 7-4 shows a similar R&S tasking matrix with its horizontal lines numbered and its vertical columns identified by letters. Use this system if you wish to modify only one specific asset tasking on the matrix. For example:

- Line 3C - 8.

- Line 3D - AB434160.

- Line 3E - Refer to 7E.

In this example, you have just told Task Force 1-10 to establish an OP at an NAI. The OP is to observe an alternate position for an MRC at

NAI 8. You updated the mission of one asset without reconstructing the entire matrix.

These are just two examples of a technique you can use to quickly retask your deployed R&S assets. There are many more. The key is to establish a standard to quickly and easily modify your R&S plan based on your commander's changing needs.

A PRIORITY	B NAI	C START	STOP	D SIR/INSTRUCTIONS	E A Co	F B Co	G C Co	H D Co	I E Co	J SCOUTS	K MORTARS	L GSR	M REMBASS	N	O	P COORDINATION	Q REPORTS
1	1 2 3	1600		When, where, and what type equipment does the recon element have? Do not engage.						x						With B Co for route to screen position. With GSR team and REMBASS team.	Per bn R&S SOP.
1	2 3	1800		How is enemy infiltrating? On foot or vehicle? Speed and direction of movement?									x			With B Co for passage. With scout platoon and GSR team.	Report on intel net when not collocated with TOC. Report every 45 minutes from 1800-2000, 2300-0230, and 0400-0630. Other hours see SOI for alternate frequency.
1	2 3	1700		Same as above. Establish alternate positions to answer same.								x				With A and B Co for passage to positions. With scout platoon and REMBASS team. For alternate sites, coordinate with B and C Co.	Report on intel net. Negative reports required hourly. See SOI for alternate frequency.
2	6	1800		How is enemy reconnoitering the southern flank? Strength and type vehicles? Will enemy attack southern flank?			x									With scout platoon and GSR team.	Per bn R&S SOP and tactical SOP.
2	4	1830 to 0600		Conduct zone recon in zone Duck. How is enemy reconnoitering northern flank? Strength and equipment? Will enemy attack northern flank?					x							With scout platoon and GSR team.	Per bn R&S SOP and tactical SOP.
3	5	1900 to 0600		Conduct security and recon patrols along route Water. Motorized rifle troops likely infiltrating.			x									With scout platoon and GSR team.	Per bn R&S SOP and tactical SOP.

Figure 7-3. Example of a modified matrix.

85

	A	B	C	D	E	F
	UNIT TASKING	PRIORITY	NAI	LOCATION	REPORTING REQUIREMENT EVENT OR INDICATOR	REMARKS
1	TASK FORCE 1-10 Recon axis speed.	1		See R&S overlay	Conditions that affect trafficability and maneuver ability. Obstacles: Type, size, and orientation.	Report as obtained.
2	Recon.		2 2A	AB474155 AB466136	2 to 3 x BMP2's, 1 x T45B; possible obstacle (single strand wire concertina).	Possible combat security outpost. Report NLT 010100Z SEPXX.
3	Establish OP.		6A	AB427185	Surveillance of activities on OBJ CAT	Establish position NLT 312200Z. AUGXX.
4	Recon.		4 4B 4A	AB453165 AB430145 AB453138	3 x MRP's with 7 to 8 x BMP2's, 2 to 3 x T64B's in prepared positions. Main obstacle array is from 800 to 1,000 m forward of MRC position.	Report all fighting positions. Report obstacle type, size, and orientation NLT 101200Z SEPXX.
5	Recon.		6	AB434160	2 to 3 x AT5(BRDM) systems or 2 to 3 x T64Bs.	Possible MRB reserve. Report as obtained.
6	Recon.		8	AB410158	3 x MRP's with 7 to 8 x BMP2's in prepared fighting positions. Obstacle array 800 to 1,000 m forward of MRC positions.	Report all fighting positions. Report obstacle type, size, and orientation NLT 011200Z SEPXX.
7			8A	AB450103	Alternate position for MRC at NAI 8.	

Figure 7-4. Modifying an R&S matrix.

CHAPTER 8

Augmenting or Task Organizing Reconnaissance and Surveillance Missions

Data gathered from different training exercises and the training centers indicate maneuver battalions typically overuse the scouts. Very often the scout platoon is the only R&S asset actively collecting on the battlefield. This usually results in a dead scout platoon, and many unanswered PIR. To increase the effectiveness of the scout platoon, other R&S assets, and the overall collection capability, you should augment or task organize as many R&S missions as possible.

Augmenting and task organizing are two different concepts that strive for the same end result. In this field manual, the term "augmenting" is used to describe using numerous assets at the same time to support the R&S plan. You task with independent or dual R&S missions to augment the R&S effort; units are not augmented.

There are circumstances in which you do not want to place R&S assets under C² of some of your subordinate units; you or the S3 want to maintain control of these assets to task or move them quickly without disrupting the other units.

There may be times when you or the S3 want R&S assets under the C² of subordinate units. Both of these concepts—augmenting and task organizing—can be used to implement productive R&S plans. The following examples provide augmented or task organized R&S missions and what each accomplishes.

Task Organized with Engineers and Artillery Forward Observers Attached to Reconnaissance Patrol

In this example you want to deny or confirm enemy activity at NAI 1 and to check on obstacles and booby traps along the road. You also want to see if the enemy has begun to prepare a possible fording site at NAI 2. Based on key intelligence you have provided, the S3 decides to send a mounted patrol for this mission. To increase mission effectiveness, the S3 also attaches some engineers and artillery FOs. Figure 8-1

87

is a sample R&S task organization with engineers and artillery FOs attached to reconnaissance patrol.

The engineers will be able to inform you of tampered-with obstacles and adjusted roadside booby traps. The FOs will be able to call for indirect fire if the patrol finds any prepared fording or bridging sites along the river. In this example you have three elements, each complementing the overall effectiveness of the reconnaissance mission. The reconnaissance element must receive instructions to clear, mark, breach, and/or bypass instructions before mission execution.

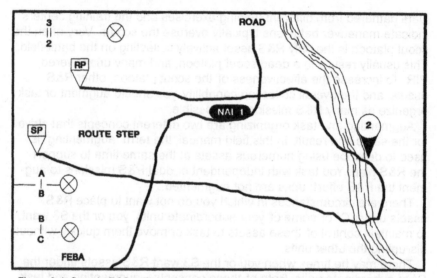

Figure 8-1. Sample R&S task organization with engineers and artillery FOs attached to reconnaissance patrol.

Task Organized with Signal Assets, Observation Post, and Forward Observers Attached to Extended Reconnaissance Patrol

Figure 8-2 is a sample task organization with signal assets, OPs, and FOs attached to extended reconnaissance patrol. In this example you have a reconnaissance patrol conducting a zone reconnaissance in Zone Buck. Due to the extended range of the patrol and the terrain, radio communications will not reach from Zone Buck to the TOC. You also have two NAI, 4 and 7, needing surveillance during a particular time window. NAI 4 is a high speed avenue of approach exiting the battalion to the north. AI and NAI 7 is a road intersection. To augment this patrol you have a signal retransmission element, OP, and artillery FOs attached to the patrol.

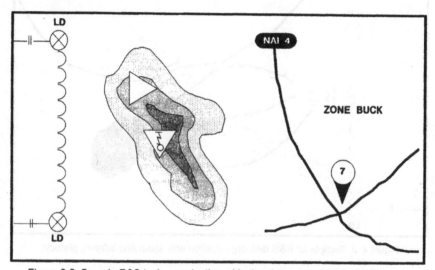

Figure 8-2. Sample R&S task organization with signal assets, and FOs attached to extended reconnaissance patrol.

The retransmission element enables the reconnaissance patrol to report what is in Zone Buck; the OP and FO enable you to have surveillance of the NAI; and the FO allows you to interdict any targets moving along the NAI or the roads.

Scouts with Infantry

In this example your battalion is preparing to conduct a raid on Objective Bear. The terrain along the river consists of thick forests. You only have two hours to reconnoiter from the LD to the objective and to mark infiltration routes. You and the S3 decide to task organize this mission with the scout platoon and two infantry platoons. Figure 8-3 is a sample of R&S task organization with scout and infantry platoons.

The scout platoon leader is the reconnaissance commander for this particular mission. The battalion has cross-trained one of the infantry platoons as the backup scout platoon. The scout platoon and the backup platoon mark crossing sites and infiltration routes while the other infantry platoon provides security. Once the scout platoon has marked the routes, they dismount and reconnoiter Objective Bear while the two infantry platoons provide security.

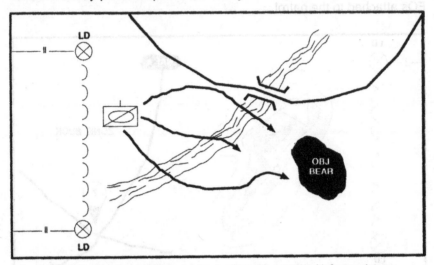

Figure 8-3. Sample of R&S task organization with scout and infantry platoon.

D Company, Scout Platoon, and Ground Surveillance Radar Effort Augmented

Figure 8-4 is a sample of an augmented R&S platoon mission. In this example your battalion is in the defense and has tasked Company D to conduct a route reconnaissance and provide surveillance of NAI 3 for four hours. The scout platoon is conducting a screen in the north. Two GSRs are providing surveillance of the flanks. Each asset is conducting an independent R&S effort. This particular mission concept allows you or the S3 to move or assign additional missions to these assets with minimal disruption to the entire battalion.

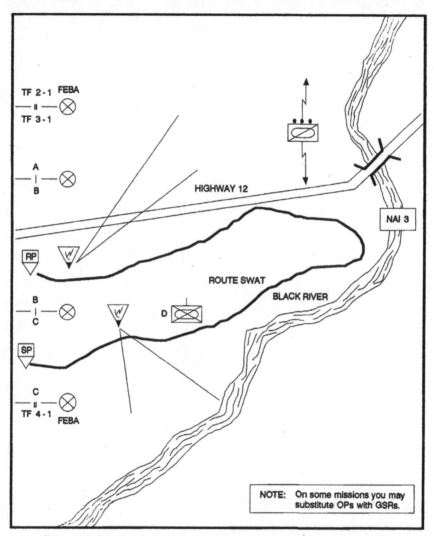

Figure 8-4. Sample augmented with D Company, Scout Platoon, and GSRs.

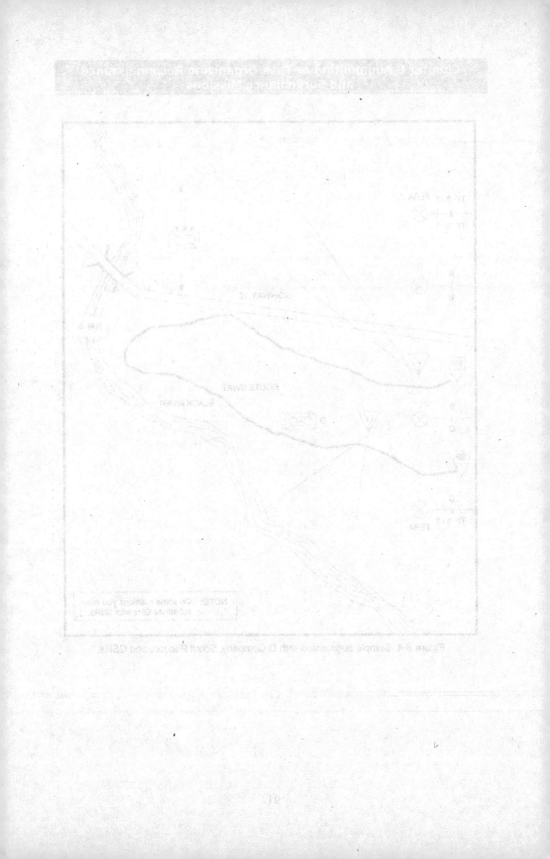

Figure 8-4. Sample sand table with D Company, Scout Platoon, and GSR's.

CHAPTER 9

Reconnaissance and Surveillance in Offensive Operations

Planning R&S missions in offensive operations requires close coordination between the S2, S3, FSO, and ALO. The chance of fratricide multiplies during these operations. The reason is R&S assets are usually conducting missions as the friendly units maneuver through these areas and engage possible enemy targets with direct or indirect fires.

Detailed reconnaissance is the initial requirement for offensive operations. Due to the limited time available to conduct detailed reconnaissance, it is imperative you use, with augmentation, all available reconnaissance assets. Recent training exercises reveal that many times S2s do not construct complete R&S plans for offensive operations. To make sure you construct complete R&S plans, consider three general areas when planning for offensive operations:

- Detailed reconnaissance.

- Surveillance of the objective.

- Ongoing R&S planning.

Detailed Reconnaissance

The first area of planning consideration stresses a detailed reconnaissance from the LD/LC to beyond the objective. During this phase you need to plan missions which answer the PIR and provide the commander and S3 with detailed information about the terrain and enemy that lie between them and the objective.

You should have assets first complete those missions designed to provide specific information that will answer PIR, so gathered information can reach the TOC in time for the commander and the S3 to make any changes to COAs or to finalize the OPORD. There are basically two areas in which to conduct this detailed reconnaissance: along the friendly AAs and at the objective.

93

Reconnaissance Along the Avenues of Approach

Typical reconnaissance missions along the AA are to:

- Detect, pinpoint, classify, and report location, dimension, and type of all obstacles (constructed or natural).

- Detect gaps or bypasses of obstacles.

- Provide surveillance and security of marks, gaps, breaches, and bypasses of obstacles.

- Report trafficability along AA.

- Establish OPs overlooking AA.

- Reconnoiter terrain and suspected enemy locations capable of overmatching and placing effective fire on the AA.

- Detect locations and strength of enemy R&S assets along the AA.

Reconnaissance of the Objective

Typical reconnaissance missions in the area of the objective are:

- Pinpoint fighting positions. Provide strength, weapon orientations, and description of fighting positions.

- Detect obstacles and prepare to mark. Detect breaches, gaps, and bypasses.

- Reconnoiter area around the objective (area depends on METT-T) to detect possible reinforcements or counterattack elements.

- Establish OPs to maintain surveillance of the objective.

As reconnaissance assets conduct these missions, you must ensure security, surveillance, and CR assets are providing coverage to the maneuver elements while they are preparing for this offensive operation. It appears to be two distinct R&S missions taking place at the same time. The first mission is providing support to the units that are preparing for the offensive OPORD. The second mission conducts R&S to answer PIR concerning the actual offensive operation.

94

Surveillance of the Objective

Now it is time to focus on the second area of planning consideration which stresses surveillance. Surveillance focuses on:

■ The objective.

■ Terrain along the friendly avenue of approach (AA).

■ Possible enemy reinforcement routes.

During this phase, you must make sure the S3 and all maneuver elements know the locations of surveillance assets.

Surveillance of the objective should detect any changes while the friendly elements are maneuvering forward. The surveillance assets report any enemy leaving or entering the objective area.

Any terrain that can control the friendly AA should be covered with surveillance or controlled by one of the seven BOS listed at the bottom of the DST.

Any routes leading into the friendly AA or the objective should be covered with surveillance to provide early warning of an enemy counterattack or reinforcements approaching. Again, these surveillance operations occur while the friendly elements are maneuvering toward the objective.

Ongoing Reconnaissance and Surveillance Planning

The third area of planning consideration focuses on both R&S missions. This area concerns planning R&S missions once your unit takes the objective. The S3 can now task subsequent R&S missions to those assets which provided surveillance to the maneuver elements. These R&S missions depend on the type of follow-on missions planned.

If the unit's mission is to reconsolidate and prepare to continue the attack to a subsequent objective, you should have a plan to continue R&S activities forward to the next objective. Remember, planning is continuous. After you accomplish this, your unit can implement missions discussed in supporting the second and third areas of planning considerations. A key scout mission is maintaining visual contact with the enemy.

If the unit's mission is to occupy and defend the objective, you should recommend an R&S plan stressing early warning and CR operations.

If the unit's mission is to pursue the fleeing enemy, you could recommend that scout elements provide flank security as other maneuver elements conduct guard operations.

The most important aspect of the final planning consideration is that it be planned out well in advance. This ensures the assets are prepared to execute the mission, not reorganizing the objective.

The three areas of planning considerations previously stated work particularly well in a deliberate attack. You can apply these same principles for a movement to contact.

Do not be misled into thinking these three areas of planning considerations take place independent of each other at different times. On the contrary, many times these missions overlap.

We have shown you a technique for constructing complete R&S plans in offensive operations. Refer to Chapter 12 for examples.

CHAPTER 10

Intelligence Support to Counterreconnaissance

The S3 is in charge of the CR mission. However, the S2 plays a critical role in developing the battlefield situation in enough detail to allow the S3 to target, destroy, or suppress the enemy's R&S assets.

Staff Officers

A number of staff officers participate in CR planning and execution. Essentially, you must find the enemy's reconnaissance units before they can find and report back on friendly unit locations. You must process information quickly and pass targeting data to the S3 and the FSO. Those staff officers involved in CR planning and execution are discussed below.

Intelligence Officer

The S2 must be knowledgeable about the enemy, weather, and terrain. Using this knowledge, the S2:

- Identifies enemy reconnaissance HVTs.

- Recommends engagement areas and ambush sites (TAI).

- Recommends HPTs.

- Makes sure electronic warfare support measures (ESM) support any planned use of EW against enemy reconnaissance elements.

- Develops an R&S plan to find enemy reconnaissance well forward.

- Evaluates vulnerability of R&S assets to enemy R&S and target acquisition capabilities.

Operations and Training Officer

The S3:

- Integrates fire, maneuver, obscurants, and EW to destroy or suppress enemy reconnaissance.

- Task-organizes the unit to defeat enemy reconnaissance well forward.

- Plans use of EW to suppress enemy reconnaissance.

- Determines and plans for use of engagement areas and ambush sites (TAI).

- Determines HPT based on the commander's intent and input from the S2 and the FSO.

- Develops deception plans to deceive enemy reconnaissance.

- Develops, executes, and monitors the unit OPSEC program.

- Uses targeting data from the S2.

Fire Support Officer

The FSO:

- Plans and coordinates all indirect lethal and nonlethal means to destroy or suppress enemy reconnaissance.

- Provides appropriate fire support coordination measures to protect the R&S participants (such as no fire areas or restricted fire areas).

- Uses targeting data based on FOS and organic or supporting target acquisition radars.

- With the S2, recommends HPTs and TAI.

- Needs targeting data from the S2 and also specific weather and terrain data for targeting and weapon emplacement.

IEWSE Officer

The IEWSE officer:

- Recommends the use of MI battalion assets, if attached or in DS.

- Informs the commander, S2, and S3 of the status and location of MI battalion assets within the unit's AO.

- Acts as liaison between the maneuver unit and the MI battalion S3.

- Provides expertise on EW planning and use.

- Receives priorities from the S3 and ESM priorities from the S2.

Air Defense Artillery Officer

The ADA officer:

- Provides early warning of enemy fixed-wing attack aircraft and rotary-wing aircraft.

- Plans for and provides air defense coverage of friendly units well forward.

- Recommends the use of ADA assets.

- Needs information on the terrain and weather from the S2 to place assets.

- Receives information on the air threat from the S2.

- Needs ADA priorities and weapons status from the S3.

Engineer Officer

The engineer officer:

- Recommends the placement and types of obstacles to stop or slow down enemy reconnaissance.

- Provides the S2 with information on the state of the terrain and enemy special reconnaissance activities.

- Receives information on the terrain, weather, and enemy from the S2.

- Needs engineer priorities and the unit scheme of maneuver from the S3.

Chemical Officer

The chemical officer:

- Monitors contaminated areas.

- Plans the use of obscurants to suppress enemy reconnaissance.

- Provides expertise on areas of likely enemy NBC use.

- Receives precise weather data.

Several primary and special staff officers can provide you with information; however, they also require information from you. Remember, you are an integral part of the targeting process. You recommend where to best engage enemy reconnaissance units. You also recommend which enemy reconnaissance elements are the most important for your unit to destroy or suppress (such as HPTs). This implies close coordination and synchronization among the S2, IEWSE, S3, FSO, and the rest of the staff.

Mission Planning

To plan the CR mission, you should know something about how terrain and weather will affect reconnaissance operations. You should also

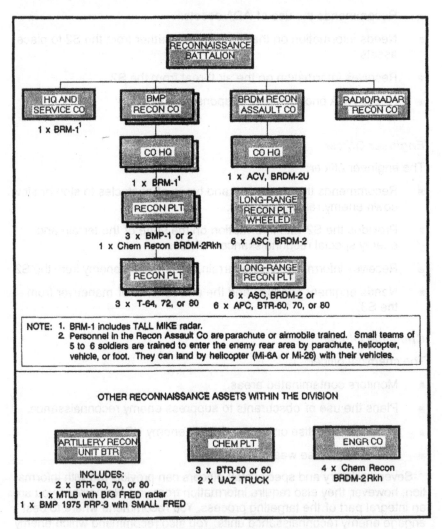

NOTE: 1. BRM-1 includes TALL MIKE radar.
2. Personnel in the Recon Assault Co are parachute or airmobile trained. Small teams of 5 to 6 soldiers are trained to enter the enemy rear area by parachute, helicopter, vehicle, or foot. They can land by helicopter (Mi-6A or Mi-26) with their vehicles.

OTHER RECONNAISSANCE ASSETS WITHIN THE DIVISION

Figure 10-1. Division reconnaissance assets.

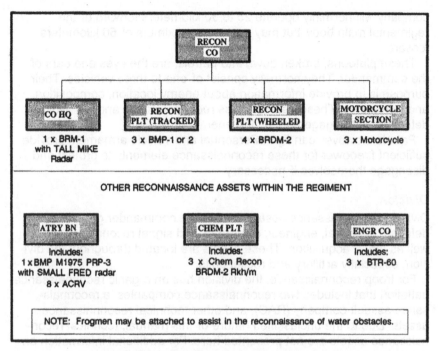

OTHER RECONNAISSANCE ASSETS WITHIN THE REGIMENT

NOTE: Frogmen may be attached to assist in the reconnaissance of water obstacles.

Figure 10-2. Regimental reconnaissance assets.

know threat reconnaissance operations, equipment, doctrine, and tactics. Remember, seek the assistance of your CI team when evaluating enemy capabilities.

Each threat division and regiment has organic ground reconnaissance units. These units either confirm information from other systems or develop their own information. They gather information primarily by patrolling and avoiding contact. Reconnaissance units may conduct raids or ambushes to gather information. Figure 10-1 shows division reconnaissance assets. Figure 10-2 shows regimental reconnaissance assets.

Reconnaissance patrols will usually be reinforced with tanks or additional armored personnel carriers (APCs) from the lead units. In addition, the lead battalions may use reinforced platoons as combat reconnaissance patrols.

Regiment

A reconnaissance company with two reconnaissance platoons provides regimental reconnaissance. These normally mounted platoons perform reconnaissance across the regimental front. The reconnaissance company will normally operate 25 to 30 kilometers forward of the regimental main body, but may operate a maximum of 50 kilometers forward.

These platoons, broken down into patrols, are the eyes and ears of the commander. They normally consist of one to three vehicles. Their purpose is to provide information about enemy location, composition, and formations. These patrols stress reconnaissance and will avoid detection and engagement by the enemy.

Patrols, however, can fight. Personnel and vehicle armament provide sufficient firepower for these reconnaissance elements to protect and disengage themselves if necessary.

Division

Division reconnaissance assets provide the commander ground, air defense, chemical, engineer, electronic, and signal reconnaissance, as well as target acquisition. These assets are located throughout the division, especially artillery and rocket units.

For troop reconnaissance, the division has an organic reconnaissance battalion that includes two reconnaissance companies, a reconnaissance assault company (RAC), and other technical reconnaissance assets. Due to the unclassified nature of this manual, technical reconnaissance assets are not presented here. For additional information on the technical reconnaissance assets, refer to the Defense Intelligence Agency (DIA) Study, "Reconnaissance and Surveillance and Target Acquisition of the USSR."

The two division reconnaissance companies will normally provide coverage across the division front, operating between the regimental reconnaissance company and RAC. These companies typically perform close reconnaissance missions for the division commander, with a primary mission of reconnaissance rather than combat.

Ideally, these companies will locate high priority targets, such as headquarters and C^3 facilities, as well as unit deployments and movements. Normally, these units will operate as small patrols of two to three vehicles with troops mounted. Troops will dismount to perform foot patrols or ambushes to gather information. However, their vehicles will not be far away.

The RAC (also called long-range reconnaissance company) performs division long-range reconnaissance. It also provides the division commander with a look-deep capability out to 100 kilometers. Small teams of five or six soldiers from this company can be inserted by parachute, helicopter, vehicle, or on foot to collect information within the enemy rear area. These teams will move primarily on foot, avoiding engagements with enemy forces, and will locate high priority targets within the enemy's division rear and corps forward area.

While the primary mission of these troops is reconnaissance, they may also have secondary missions to conduct disruptive operations in the rear area, such as:

■ Ambushes.

■ Prisoner snatches.

■ Traffic diversions.

■ Disruption of lines of communication (LOC).

■ Limited attacks against important targets of opportunity.

When not operating in the enemy area, this company is capable of providing additional reconnaissance patrols mounted in their organic vehicles within the division area.

Reconnaissance Fundamentals

Reconnaissance plays an important part in the overall intelligence-gathering system. It can provide confirmation of other collection assets. It often provides initial information that can be confirmed by other means, such as electronic or signal reconnaissance.

Troop Reconnaissance

Troop reconnaissance is responsive to the commander's needs and can provide timely information on which to base command decisions.

Division and regimental reconnaissance efforts are carefully planned, coordinated, and supervised by the chief of reconnaissance, while battalion and lower commanders must accomplish the task themselves.

Reconnaissance Patrols

Reconnaissance patrols will gain information by observation; they will bypass defenders. However, they will fight if required. Normally the tanks and BMPs will overwatch the BRDMs. BMPs and BRDMs will make a detailed reconnaissance of all likely enemy positions, with the tanks providing cover.

Chemical-Engineer Reconnaissance

Chemical-engineer reconnaissance teams will move behind the lead reconnaissance elements. When obstacles or contaminated areas are located, they will be marked and their locations reported to the regimental commander. Reconnaissance elements will use bounding overwatch techniques. Figure 10-3 shows the Soviet reconnaissance overwatch. Figure 10-4 shows Soviet reconnaissance overwatch with patrols. Figure 10-5 shows Soviet technique patrols with overwatch.

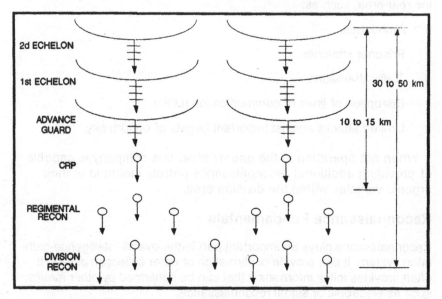

Figure 10-3. Soviet doctrinal deployment (meeting engagement).

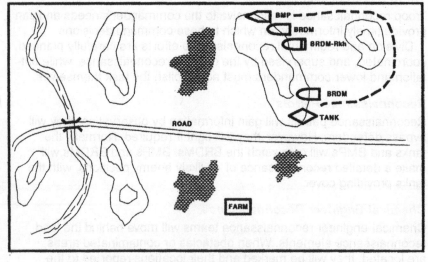

Figure 10-4. Soviet reconnaissance overwatch.

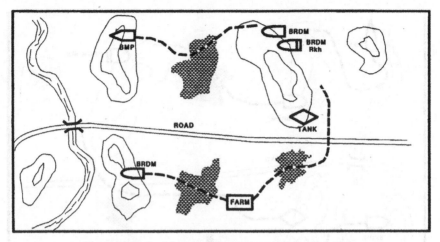

Figure 10-5. Soviet reconnaissance overwatch with patrols.

Using Intelligence Preparation of the Battlefield to Support Your Counterreconnaissance Effort

Once again, the IPB process can help in your planning. The two most important products you will develop in CR are situation templates and event templates. It will be helpful to develop a series of situation templates which depict enemy reconnaissance movement. Such templates allow you to develop your event template and let your S3 visualize how you expect the enemy to conduct their reconnaissance battle. Figure 10-6 is a sample of one such situation template.

Situation Template

Do not make the mistake of thinking the enemy's reconnaissance will use the same AAs as the enemy main force. Remember, enemy reconnaissance elements will most likely operate as two or three vehicles. Such small elements can traverse almost any kind of terrain. Keep in mind, the mission of reconnaissance is to seek and report information, not to fight. Therefore, enemy reconnaissance will use routes that have plenty of concealment and cover.

Also remember, enemy reconnaissance is looking for the best route of attack; the enemy may decide that attacking over rough terrain is preferable to attacking open, but heavily defended, country. For this reason, be sure to consider your entire AI when you develop

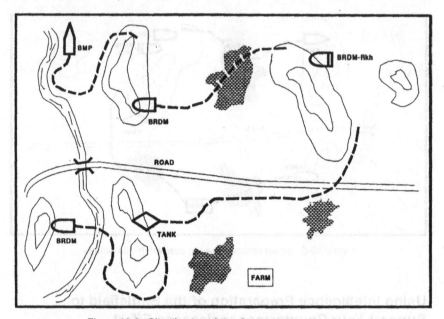

Figure 10-6. Situation template of enemy reconnaissance.

your situation templates. Do not get "tunnel vision" and consider only obvious AAs or MCs.

Look at all ways the enemy can enter your sector, including using No-Go terrain! As a general rule, the more concealment or protection a route provides, the more likely it will be used by reconnaissance elements. The outcome of well prepared situation templates is a commander and staff that have a good indication of what the enemy will look like on the battlefield. This eventually will save your R&S assets many hours of unnecessary reconnaissance or surveillance.

Event Template

Based on your situation templates, develop your event template. Your event template will show where on the battlefield you expect to see enemy reconnaissance elements. Then concentrate your R&S attention on those areas (NAI) to detect enemy reconnaissance activity.

Remember, the key to CR intelligence support is finding those enemy reconnaissance units before they can discover friendly positions and report back. Therefore, you must carefully study the effects of weather and terrain on enemy reconnaissance to determine at what point the enemy can observe friendly positions. Usually, this is a function of observation (LOS) and visibility in your unit's AI.

Compare these limits with the enemy's known reconnaissance observation capabilities (such as infrared, thermal, light enhancement, and telescopic). As you do this, you will begin to identify a limit of enemy advance (LOEA). Essentially, you must prevent the enemy from going beyond this limit because past that limit, the enemy can observe friendly positions.

Figure 10-7 shows an example of an LOEA, or you can recommend a phase line (PL) that represents the LOEA.

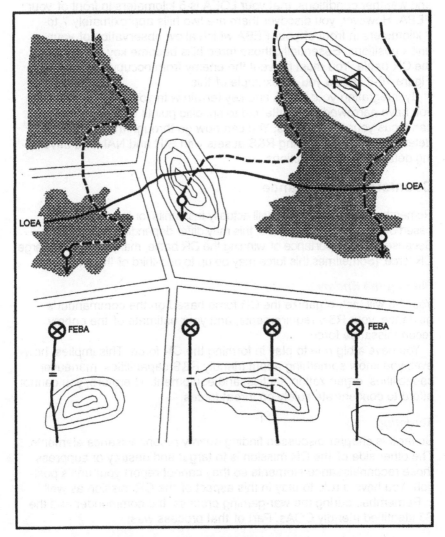

Figure 10-7. Limit of enemy advance.

R&S Plan

You should focus your R&S assets forward of the LOEA to identify enemy reconnaissance before they can spot your unit's positions. Additionally, your analysis of the terrain may indicate there are isolated terrain features forward of the LOEA you must control to prevent enemy observation of your unit's position.

For example, you may have determined, based on general terrain and weather conditions, that your LOEA is 5 kilometers in front of your FEBA. However, you discover there are two hills approximately 7 to 8 kilometers in front of your FEBA which allow observation of your unit's positions. Essentially, those three hills become key terrain for the CR battle. You must prevent the enemy from occupying those hills. Figure 10-8 shows you an example of this.

By integrating your LOEA and key terrain with your event template, you have narrowed the battlefield to specific points or areas where you can focus your R&S assets. You can now go through the process of determining SIR, matching R&S assets with SIR and NAI, and developing detailed R&S instructions.

Counterreconnaissance

Remember, normally your S3 will actually task units for the CR mission based on your input. (Of course, this may differ depending on unit SOP.) Because of the importance of winning the CR battle, many units use a large CR force. (Sometimes this force may be up to one-third of the entire unit.)

Finding the Enemy

Your S3 will task-organize the CR force based on the commander's guidance, your R&S requirements, and your estimate of the enemy reconnaissance force.

You have a big role to play in forming the CR force. This implies, however, you know something about friendly R&S capabilities, maneuver capabilities, organization, tactics, and equipment. Therefore, you cannot afford to concentrate solely on threat forces.

Targeting

So far this chapter discussed finding enemy reconnaissance elements. The other side of the CR mission is to target and destroy or suppress those reconnaissance elements so they cannot report your unit's position. You have a role to play in this aspect of the CR mission as well.

Remember, during the war-gaming process, the commander and the S3 identified friendly COAs. Part of that process was:

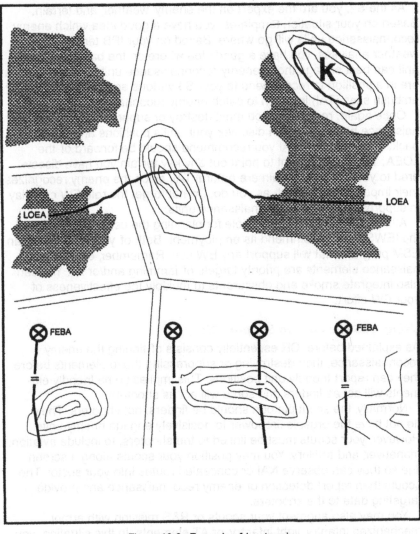

Figure 10-8. Example of key terrain.

- Developing TAI.

- Deciding how best to engage enemy units at TAI.

- Formulating decision points or lines.

As the S2, you are the expert on the enemy, weather, and terrain. Based on your situation templates, you have a good idea which enemy reconnaissance units will go where. Based on your IPB terrain and weather analyses, you have a good idea where on the battlefield your unit can best engage those enemy reconnaissance units. Therefore, you are in a position to recommend to your S3 various engagement areas or ambush sites (TAI) in which to catch enemy reconnaissance elements.

Once again, remember, you must destroy or suppress enemy reconnaissance before they can discover your unit's positions and report back. Therefore, any TAI you recommend should be forward of the LOEA. You will also want to point out any key terrain you have discovered to your S3. Key terrain are natural TAI, since the enemy recognizes their importance as much as you do. Do not forget the role EW can play in suppressing enemy reconnaissance units.

Although your S3 is responsible for planning the use of EW, you and the IEWSE can recommend its employment. Both of you should plan an ESM program that will support any EW use. Remember, enemy reconnaissance elements are priority targets of jamming and/or DF. You can also integrate smoke and obscurants to multiply the effectiveness of your EW effort.

Using R&S Missions to Support CR

As explained before, CR essentially consists of finding the enemy reconnaissance, then destroying or suppressing those elements before they can report friendly unit positions. This implies some friendly elements will act as finders and some will act as shooters.

Normally, the scout platoon should be finders, not shooters. They do not have the organic firepower to decisively engage enemy units. However, your scouts must be linked to the shooters, to include aviation, maneuver, and artillery. You may position your scouts along a screen line so they can observe NAI or concealed routes into your sector. The scouts then report detection of enemy reconnaissance and provide targeting data to the shooters.

You may also augment your scouts or R&S mission with armor, mechanized infantry, light infantry, or AT elements. In this situation, you might employ your scouts as roving teams. The scout element finds the enemy reconnaissance, informs the S3, who then calls in the armor, infantry, aviation, or indirect fire assets to destroy it. Figure 10-9 is an example of scout employment to screen concealed routes. Figure 10-10 is an example of the use of mechanized infantry with scouts under operational control.

To effectively plan your portion of the CR mission, you need to know how threat reconnaissance operates. For additional information on threat smoke and obscurant employment, refer to the Joint Test Command Group manual, 61 JTCG/ME-87-10, Handbook for Operational Testing of Electro-optical Systems in Battlefield Obscurants.

This chapter discussed various staff roles in the CR mission, how you can contribute, and how the threat performs reconnaissance at regimental and division levels. It is also important for you to do your "homework" to find out how the various threat forces conduct dismounted reconnaissance.

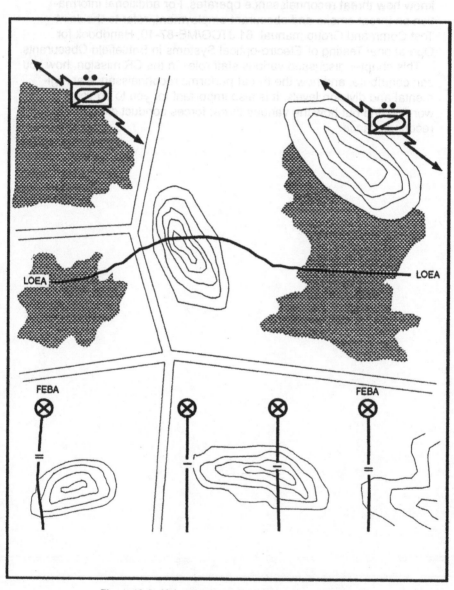

Figure 10-9. Using scouts to screen concealed routes.

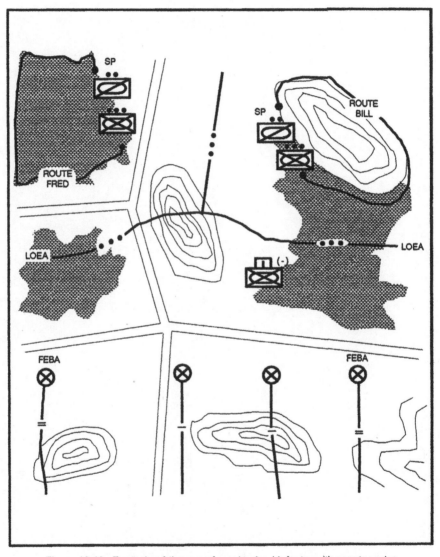

Figure 10-10. Example of the use of mechanized infantry with scouts under operational control.

CHAPTER 11

Reconnaissance and Surveillance in Low-Intensity Conflict

During LIC operations, R&S must provide your commander a wide range of information in a complex environment.

Factors

Factors to consider when planning R&S in an LIC environment include:

- U.S. forces mission—counterinsurgency, combatting terrorism, peacekeeping operation, or peacetime contingency operations.

- Threat—conventional forces, insurgent forces, terrorists, demonstrators, or a combination of two or more.

- Environment—social, psychological, political, and economic factors. Terrain and weather are also important considerations.

- Host nation government—support, information sharing, security forces, and military forces.

The U.S. force's mission, environment, and host-nation government are influences that have an affect on what we do offensively or defensively. However, the LIC threat will require you to spend the most time learning how to predict the enemy's next move.

Threat

The LIC threat can range from demonstrations, terrorist acts, insurgent or guerrilla activity to confrontations with conventional forces. The characteristics of a threat force depend on the level of insurgency. U.S. maneuver forces will most likely face insurgent forces or a hostile country conventional military force. Your unit may face demonstrators or terrorist threats. But primary population control responsibility is with the host nation.

Insurgent or guerrilla forces usually fight in small cells. They carry light weapons and can concentrate forces against major facilities, then disperse after the operation. Insurgent forces can operate in urban

115

areas but prefer remote areas for better concealment and security. You can expect to fight squad- to platoon-size forces when facing insurgent forces. They will rely on:

- Well-planned ambushes.

- Attacks on soft targets.

- Sniper and mortar attacks.

The objective is to demoralize and frustrate their opponent by attacking a variety of targets in a wide AO. Segments of the populace can play a key role in the insurgent intelligence net; in which case, they would become a primary target of friendly CI efforts.

Guerrilla forces need support from political sympathizers or foreign powers. They need an effective system of obtaining food, ammunition, weapons, equipment, and training. In some cases insurgents conduct raids for equipment. By eliminating insurgent supply nets and sources they lose combat effectiveness.

See DA Pam 381-3, How Latin American Insurgents Fight, for detailed information.

Conventional Threat Forces in LIC

Conventional threat forces in an LIC environment conduct a variety of missions. These missions involve advising and assisting insurgent forces on how to fight. Conventional threat forces train insurgents on the use of sophisticated weapons or act as leaders for insurgent units. This involvement depends on support provided by the hostile government.

Conventional threat forces can operate in traditional roles, attacking and defending to support insurgents. These forces are infantry, or mechanized infantry supported by artillery, mortars, and armored vehicles. Along with limited CAS, they could have NBC weapons.

Their equipment is a mix from several major weapons-producing countries (for example, United States, Belgium, Soviet Union, China, and West Germany). Usually this equipment is a generation or two older than that found in modern armies. However, this trend is slowly changing. The type of weapons used in an LIC environment varies from homemade weapons (mines or shotguns) to sophisticated weapons (SA-7's). Understanding the capabilities of guerrilla/insurgent weapons and collection and target acquisition systems helps you in R&S planning.

116

Guerrilla/Insurgent Operations

Guerrilla operations are those military actions executed with selected
commands and combatants. For this reason, it is necessary to obtain
specific enemy information, and to know the enemy's situation by
observation. In guerrilla operations, attacking by surprise and having
control of key terrain are essential.

Everyone who engages in guerrilla operations, besides being elusive,
must have had excellent training and preparation. The following are
general prerequisites or priorities for the preparation of an individual
guerrilla fighter:

- Physical conditioning.

- High morale.

- Individual combat training.

- Land navigation and knowledge of the terrain.

- Complete understanding of the mission.

- Clear understanding of his or her role in the mission.

- Discipline.

- Esprit de corps.

- Aggressiveness, dexterity, self-confidence, valor, and courage.

- Decisiveness and patience.

Guerrilla operations include:

- The ambush.

- The incursion.

- The surprise attack.

- Sabotage (machinery, electrical energy, and telephone).

- Infiltration (capture of personnel, weapons, and documents).

In every guerrilla operation, the execution of the mission must be
guaranteed.

Current and accurate enemy information, including terrain and weather knowledge, are key to prepare, plan, and execute the mission. Every small detail must be covered in the plan, and nothing should be overlooked.

For each guerrilla operation, training or simulated attack must be conducted and verified; these must be in terrain which closely resembles the site characteristics where the operation will be carried out.

Individual guerrilla training must be continuous; it must always strive for superiority in all aspects of training. Training must focus on the prerequisites mentioned above.

Coordination is a high priority during each guerrilla operation. Coordination ensures teamwork and helps to guarantee the success of the operation.

Selected commands and combatants, as well as weaponry, are key ingredients for the operation. Each guerrilla fighter must make full use of weapons and must not fire continuously. It is very important that strict fire discipline be followed.

Terrain knowledge, appropriate camouflage, and surprise are essential elements during the execution of a guerrilla operation.

Surprise, security, rational use of resources, and economy of force are the key principles of guerrilla warfare tactics. These must be followed in every guerrilla operation.

Guerrillas around the world typically fight the same way. They use surprise, night operations, careful planning and selection of targets, and timing to inflict the greatest damage. They are particularly sensitive to the propaganda value of the psychological impact of every action they take, from a single terrorist act (bombing or political assassination) to a major assault on a critical installation.

Guerrillas can best be described as capable, all-weather soldiers who live off the land, thus reducing the amount of rations they need to carry. They operate in their own domain and, because of their familiarity with it, can negotiate the most difficult terrain in any kind of weather, at any time of the day. They usually attack at night to ensure the element of surprise. The overall combat effectiveness of these fighters is usually good.

Upper Echelon Organization

The guerrilla's military organization is a network of insurgent groups placed in different parts of a country. It has a definite command structure based on geographical location. Thus, a "Northern Command," an "Eastern Command," a "Central Command," and a "Northwest Command" would correspond to the area of the country in which each command operates. Figure 11-1 shows a typical insurgent organization.

The guerrilla military organization is headed by a general staff with staff departments organized to fulfill training, logistic, troop, intelligence, and operational functions. Figure 11-2 shows the general staff. Figure 11-3 shows the logistic staff. Figure 11-4 shows the troop staff.

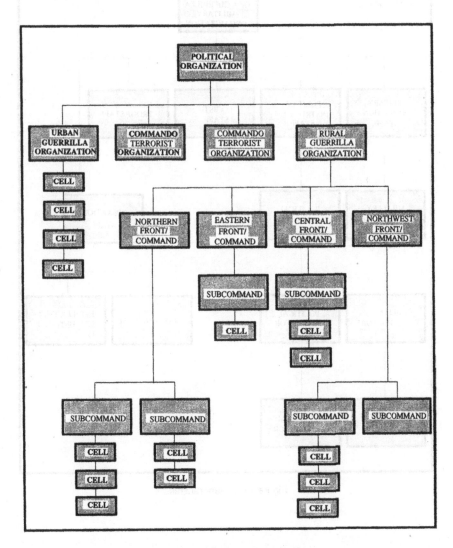

Figure 11-1. Insurgent Organization.

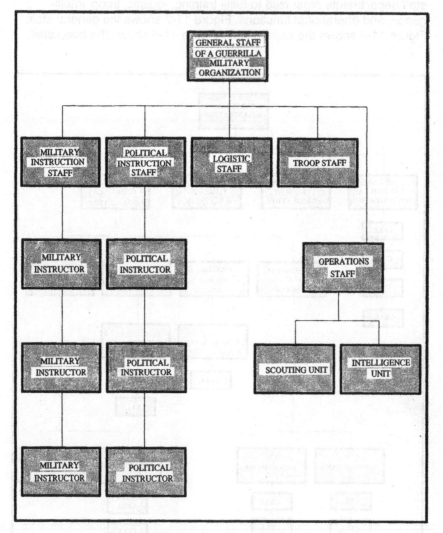

Figure 11-2. General staff.

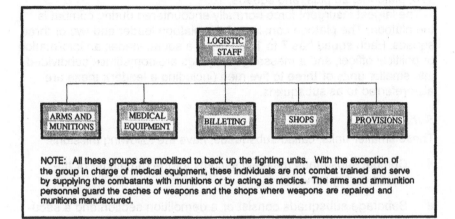

NOTE: All these groups are mobilized to back up the fighting units. With the exception of the group in charge of medical equipment, these individuals are not combat trained and serve by supplying the combatants with munitions or by acting as medics. The arms and ammunition personnel guard the caches of weapons and the shops where weapons are repaired and munitions manufactured.

Figure 11-3. Logistic staff.

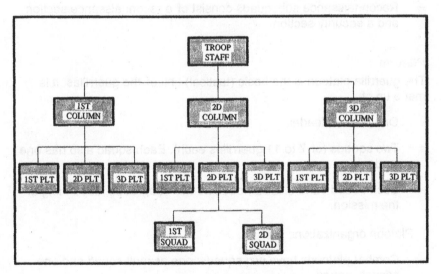

Figure 11-4. Troop staff.

The guerrilla military forces have an infrastructure ranging from a brigade of several thousand down to a cell of three to five people. Falling in between are battalions or columns with 500; detachments, 100; platoons, 20 plus; and squads, 10.

The largest insurgent force normally encountered during combat is the platoon. The platoon consists of the platoon leader and two or three squads. Each squad has 7 to 11 soldiers, a squad leader, an information or political officer, and a messenger. Squads are sometimes subdivided into smaller units of three to five men (including a leader); these are also referred to as subsquads.

Subsquads

These smaller units, called subsquads, have the following missions:

- Combat subsquads maintain security for the remainder of the squad during movement.

- Sabotage subsquads consist of a demolition section and a security section. The security section provides security to the demolition section during sabotage operations.

- Reconnaissance subsquads consist of a reconnaissance section and a security section.

Platoon

The guerrilla platoon is the basic (tactical) unit of the guerrillas. It is made up of:

- One platoon leader.

- Two squads (of 7 to 11 guerrillas each). Each squad also has one leader.

- Each squad can be subdivided into two subgroups, depending on the mission.

Platoon organizations are:

- Combat platoons have two squads: one assault squad and one security squad.

■ Reconnaissance platoons have two squads: one reconnaissance squad and one security squad.

■ Other platoons provide logistic support (storehouses for arms), withdrawal sites, communications system, transport, firing ranges and maneuver sites, underwater demolition, sabotage, and propaganda.

Command and Control

The insurgent organization and chain of command simply consists of leaders and followers. Insurgent commanders exercise control over their forces by delegating command authority down to the platoon and squad levels. Although several platoons may be committed in combat, all missions are assigned and conducted at the squad level.

Insurgent combat units may operate alone when seizing terrain, but all unit commanders must ensure that the high command provides detailed instructions for their unit. The leader of a combat unit is expected to exercise effective control of all combatants, to assign specific functions, and to see that they are strictly complied with.

Insurgents are also trained to use their own initiative when faced with unforeseen situations. The unit must be flexible and capable of solving problems associated with missions assigned by the supreme command in such areas as:

■ Communications.

■ Chain of command.

■ Mobilization of forces.

■ The use of equipment and firepower.

Equipment

Insurgents require the same combatant gear to perform their mission as the friendly forces; however, insurgents do not usually have the same equipment. Theirs is obtained from a variety of sources through a complex logistic system. Weapons, food, medicine, and other supplies are obtained by one of the following methods: black market, captured, stolen, or provided by second-party sources.

Much of the equipment is obtained from government forces through raids on isolated outposts or ambushes on military units and convoys;

additional guerrilla supplies and arms come from other countries. Some equipment is purchased abroad with money obtained through terrorist activities (kidnapping and robbery).

A more popular and practical means of obtaining military supplies is by capturing government weapons; this ensures an abundant supply of ammunition and repair parts is available. The three essential requirements of insurgent weapons are availability, simplicity, and efficiency.

Usually, each combatant has the following equipment:

■ Rifle: Belgian FAL; Israeli Galil; German G-3; Soviet AK-47 or AKM; Czech M-25; and United States M-2, M-14, or M-16.

■ Pack or knapsack.

■ Web belt.

■ Canteen.

■ Beret, cap, or hat.

■ Protective combat clothing that blends with the terrain.

■ Combat boots.

■ Knife and steel blade.

■ Weapon cleaning equipment.

■ Nylon cord, approximately 2 meters long.

■ Square of plastic, 2 by 2 meters (to protect weapon from rain).

■ Medication kit (such as aspirin, bandages)

Each unit has a radio (probably commercial type, AM or FM) to keep the insurgents informed of the news.

An insurgent unit may have one or more of the following weapons:

■ Hand grenades (fragmentation, concussion, and incendiary).

■ Grenade launchers.

■ Mortars.

■ Mines of the claymore type.

■ An assortment of AT and air defense weapons.

Artillery

Artillery is the principal fire power for some insurgent forces. Insurgents use it because of its range, volume of fire, and accuracy. The principal mission for an artillery unit is to neutralize or destroy the enemy and their means of combat. Mortars and recoilless weapons are usually the preferred artillery pieces used by the guerrilla, probably due to their mobility and portability.

Guerrillas can and will use captured heavier weapons. They will transport them by commercial vehicles into the battle area or abandon them, if necessary, if they impede their withdrawal from the area. Artillery is classified according to:

- Recoil construction and type of tube.

- Caliber: small caliber, 20 to 57 mm; medium caliber, 58 to 152 mm; large caliber, over 152 mm.

- Bore: smooth (mortar), the 205 mm has grooves.

- Firing: high angle or flat trajectory.

- Means of transport: mechanical traction or self-propelled.

- Initial velocity classified as follows: mortars from 150 to 400 meters per second; a howitzer from 300 to 600 meters per second; and cannons from 900 to 1,500 meters per second.

It is important to note that some of this equipment is homemade, such as uniforms, pistol belts, and harnesses. Insurgent camps sometimes contain factories where Molotov cocktails, booby traps, claymore type mines, grenades, and ammunition, including mortars, can be produced at little cost in a short time.

Types and Sources of Supplies

The guerrilla, by necessity, uses a wide variety of weapons, some self-manufactured, some captured, and some supplied from outside sources. In the earlier stages of a war, the weapons are usually primitive, homemade rifles, hand grenades, and claymore type mines; trails are crudely booby-trapped with Punji stakes and shallow pits lined with nail boards.

Nearly every guerrilla war has produced ingenious improvisations, both from necessity and to avoid a cumbersome logistic supply system. Nothing can be simpler to construct and use than a Molotov cocktail or a plastique bomb; and under certain conditions, nothing can be more effective.

Arms and Ammunition

All types of arms are needed for a guerrilla movement. However, there are three important factors which insurgents have to keep in mind when arms are selected: weight, range, and rate of fire. Guerrillas carry their weapons for long periods of time, thus the weapon must be as light as possible. The weapons must be effective both at short and long range. Maximum rate of fire is critical since guerrillas need to place a large amount of fire in a short amount of time. With a variety of weapons comes the need for different types of ammunition. Individuals responsible for acquiring ammunition must be able to distinguish between the different types and caliber rounds needed.

Food

Just as with arms and ammunition, food is a basic necessity for a guerrilla movement. It must be easy to carry, nutritious, and not perishable (such as chocolate, condensed milk, dried fish or meat, rice, beans, cereals, sugar, coffee). The main sources for food are local villages, supplies left behind by government troops, warehouses, stores, and the land itself.

Explosives

Explosives are the key to guerrilla operations due to their destructive power. They are used to destroy bridges, railroad lines, airports of military value, communication lines, and electrical towers. To acquire explosives, clandestine groups are formed which operate in areas where explosives are used. By attacking vehicles which transport the explosives, they are able to obtain the explosives needed.

Hand Grenades

These can be industrially or domestically manufactured. There are two types of hand grenades:

- Defensive—A metallic container that splinters; has an effective range of more than 30m and is used mostly to break out of a siege by disorganizing the enemy.

- Offensive—A container filled with an explosive charge, which when ignited, creates a proliferous blast of fire or pellets. This type is used mostly for ambushes and in attacks on garrisons.

It also serves to disorganize the enemy, as well as cause casualties.

External Supplies

Nicaragua has been the main source of external supplies since the start of insurgences in Central America. While in South America, Cuba has been the primary source guaranteeing the guerrillas a sustained rate of supply and resupply. Supplies are carried by aircraft, small boats, trucks with false bottoms, stolen buses, or pack animals. It depends on the terrain and on the control exercised by the guerrillas in the area in which they are operating.

Internal Supplies

On the local level, if the guerrillas have funds, they purchase food and medicine. However, this is rare, and these are primarily acquired by stealing and pillaging from villages and towns temporarily occupied by guerrilla groups. Known as "war taxes," farmers and merchants are threatened with death if they fail to pay (comply).

In the cities, safe houses serve as storage and distribution points for the guerrilla's supply network. Large caches of weapons and ammunition are kept at convenient, centrally located, and relatively safe geographic locations.

In regions controlled by the guerrillas, the noncombatant camp followers are required to cultivate the land for cereals and basic food grains, with the guerrillas taking half of the harvest. In some areas, sugar mills and slaughter houses are operated by collaborators of the guerrillas. Basic food stuffs such as beans, rice, cooking oil, salt, sugar, and corn are collected and stored before an insurgent offensive. When not in combat, guerrillas are able to obtain supplies with money almost anytime, anywhere.

Communications

It is impossible to direct a war without communications. Among the most important forms of communication for the guerrillas is the radio. There are two types of radio communications: tactical and operative communications and strategic communications.

Tactical and Operative

These are the radio signals used by leaders to command their units in operations, marches, and encampments. Due to the need for maneuverability and agility, radios used include walkie-talkies, citizen band, and PRC-77.

Strategic

These are radio signals used by strategic commands in order to have an overall vision of all the fronts of the war and to direct the war. In long distance communications, a variable selection of ham radios are used.

When setting up a base camp, the radio operator seeks a high location to establish effective communications. Radio waves require LOS so it is important that the radio operators set up operations at the highest point to avoid natural or artificial objects. The radio is always set upright with the antenna in a vertical position directed towards the receiver with whom the insurgent wishes to communicate. Messages are brief.

Relay stations bridge stations that do not have direct communication because of topographic obstacles or too long a distance. There are two types of relay station: manual (operated by an individual) and automatic (signal is sent out automatically when received).

Forms of Guerrilla Combat

To prepare for combat, the guerrillas must train in isolated locations. Figure 11-5 shows a typical guerrilla training complex. Before the guerrillas train on hitting targets they are indoctrinated on the typical targets they should hit. Figure 11-6 shows typical guerrilla targets.

Raid

This is a fast, surprise action carried out against an enemy position or force. Its purpose is related directly to current needs in the development of the guerrilla's campaign (arms, food, propaganda). Generally, raids are well planned and carried out in small units composed of three to twelve individuals. They occur more frequently in the initial stages of an insurgency when few individuals and arms are available. Once their goal is achieved, the force withdraws quickly and disperses.

Assault

This is a more sophisticated, complex attack designed to annihilate a target and its defenders. As assault is of a larger scale and purpose, it requires the occupation of positions and strategic locations since the guerrillas are fighting against prepared defensive positions of the enemy. At a certain phase of a local insurgency, the prime goal becomes that of eliminating enemy units, thus changing the correlation of forces in the region.

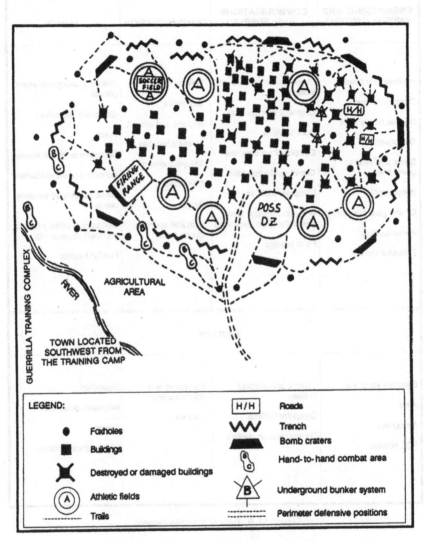

Figure 11-5. Typical guerrilla training complex.

ENGINEERING AND ENERGY SYSTEMS	COMMUNICATIONS AND SUPPLY	TRANSPORTATION	HUMAN
Hydroelectric plants	Communications lines and facilities	Railroads	Embassy and government officials
Offshore oil rigs	Chemical storage sites	Bus depot	Corporate executives
Nuclear facility sites	Dock facilities	Airports and aircraft	Police
Gas pipelines	Equipment warehouses	Trucking facilities	Dependents of the above
Dams and electric power lines	Chemical storage sites	Shipyard, docks	Schools and school buses
Water supply	Computer facilities	COMMERCE	Areas catering to personal needs
Oil and gasoline storage	Weapons storage sites (special and conventional)	Banks	Members of military forces and their dependents
Repair facilities	Food storage	Gun and sporting goods store	Foreign tourists
Explosive storage			

MILITARY			
Sensitive weapons	Logistic and storage facilities	Command and control facilities	Explosives
Arms	Computer facilities	Vehicles	Recreational facilities
Ammunition	Communication centers		Aircraft
POL storage			Maintenance facilities

Figure 11-6. Typical guerrilla targets.

Ambush

This is an action carried out by small units against a moving enemy with great superiority in soldiers and arms. Factors that influence the outcome of an ambush are location, terrain, position, camouflage, signals, and retreat. There are three types of ambushes:

■ Annihilation—enemy troops sustain the maximum number of casualties to reduce or destroy the combat effectiveness of the government forces. Figure 11-7 shows a typical annihilation ambush.

■ Harassment—enemy troops are harassed by engagement in small skirmishes to destroy their will or to distract and tire them, thus causing deterioration of morale. Figure 11-8 shows a typical harassment ambush.

■ Containment—enemy forces are surrounded by mines, obstacles, and small arms fire to halt movement to and from a specific area; usually to keep them from reinforcing a government unit in contact with insurgents. Figure 11-9 shows a typical containment ambush.

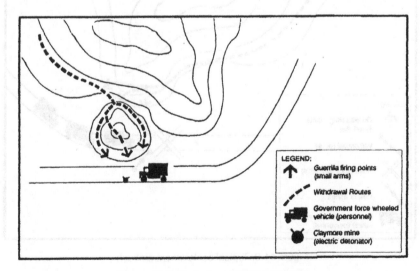

Figure 11-7. Typical annihilation ambush.

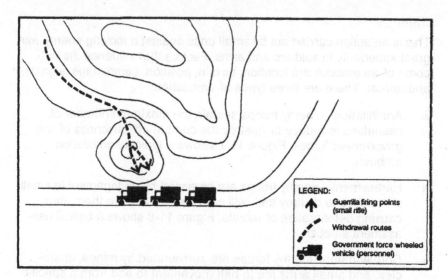

Figure 11-8. Typical harassment ambush.

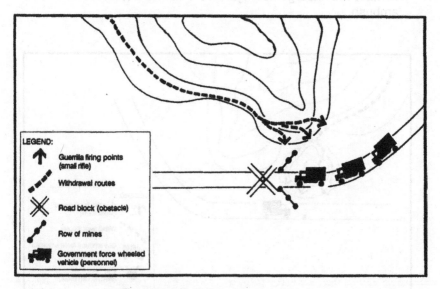

Figure 11-9. Typical containment ambush.

Insurgent Map Symbols

These are the symbols used by known insurgent groups. Figure 11-10 shows the military mapping symbols. The listing includes many military mapping symbols used by the Soviet armed forces; however, different meanings have been applied.

Insurgents normally use the symbols that are taught to them by the country that is providing training and equipment. Also refer to DA Pam 381-3 for insurgent military map symbols.

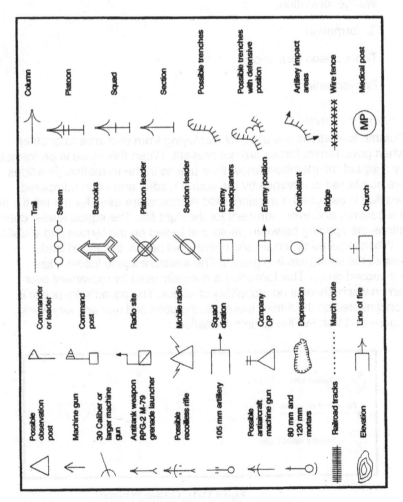

Figure 11-10. Military mapping symbols.

133

Movement Formations

Movement formations include:

- Column formation.

- Single firing or line formation.

- Diamond formation.

- Wedge formation.

- "L" formation.

- Two-echelon formation.

- Fan formation.

Column Formation

Column formations are used for deploying from one area to another when government forces are not present. When this move is performed by a squad, all members know their places in the formation. Positions are numbered in advance by the squad leader, and each numbered position is assigned a mission. Odd numbers are used for the left file of the column; and even numbers for the right file. The squad leader determines the spacing between personnel based on the terrain and visibility.

When required by terrain and operational needs, the squad leader divides the squad into two groups. The assistant squad leader takes the second group. This formation is normally used for movement over terrain where there is no probability of attack. The squad's fire power is concentrated on the flanks; therefore, the point and rear are very weak. Figure 11-11 shows the column formation.

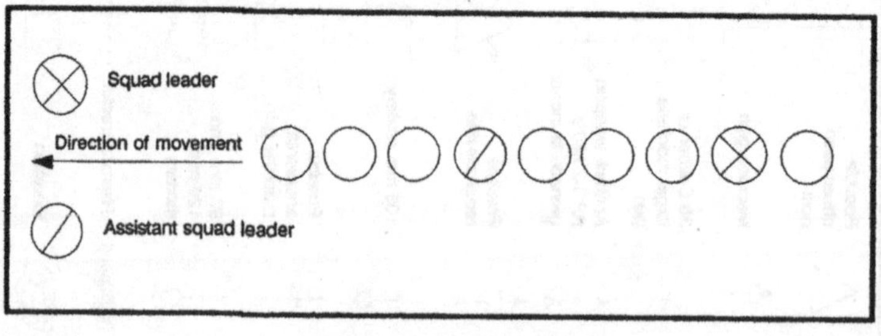

Figure 11-11. Column formation.

Single File or Firing Line Formation

This type of formation is used when necessary to cover a 100- to a 300-meter area. Figure 11-12 shows the single file or firing line formation. It is used when:

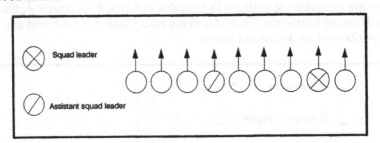

Figure 11-12. Single file or firing line formation.

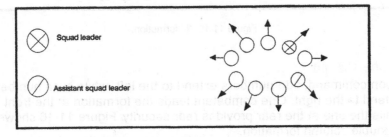

Figure 11-13. Diamond formation.

Wedge Formation

A wedge formation is used for advancing or performing reconnaissance over open terrain. This formation covers the front, as well as both flanks; however, the rear is undefended. This type of formation is used mainly to move where there is a possibility of being attacked.

It can also be used to break or penetrate an enemy barrier. Figure 11-14 shows the wedge formation.

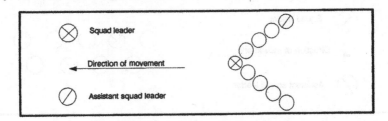

Figure 11-14. Wedge formation.

135

"L" Formation

The "L" formation is an attack formation used in two flanks. Figure 11-15 shows the "L" formation. It can be used before the assault by deploying one squad to gain a shock while the remaining squad provides security. From the formation of a single to a double column, it can quickly change to a diamond formation. These changes take place on command and are performed as discussed below.

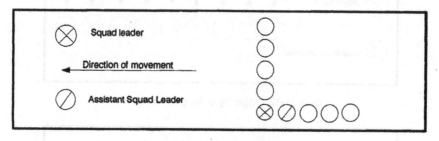

Figure 11-15. "L" formation.

Double Column

Upon command, odd numbers extend to the left, while even numbers extend to the right. One combatant leads the formation at the front while the one at the rear provides rear security. Figure 11-16 shows a double column formation.

Changing Double Column to a Diamond Formation

In order to change a double column to a diamond formation, the insurgents change the above-mentioned formation. At this time, the insurgents on the right extend to that side, while those on the left extend to the right. (See Figure 11-16.) The combatant from the rear guard at the right flank maneuvers, while providing rear security. The group on the left that heads the team secures the front.

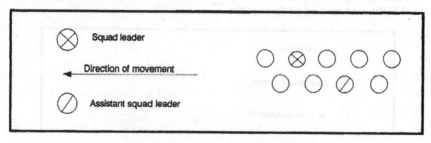

Figure 11-16. Double column formation.

Two-Echelon Formation

This type of formation is used for a deliberate attack or a movement to contact. While a squad advances, the other one supports it; upon occupying a new position, the one advancing stops and provides support while the other unit advances. Figure 11-17 shows two-echelon formation.

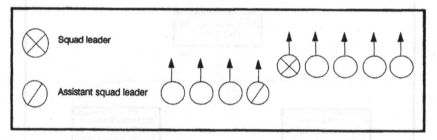

Figure 11-17. Two-echelon formation

Fan Formation

This formation is used when the commander calls the unit to assemble to assign a mission or issue other directives. The voice of command is: "SQUAD ASSEMBLE." Figure 11-18 shows the fan formation.

Squad leader

Assistant squad leader

NOTE: Assistant squad leader has no set position.

Figure 11-18. Fan shape formation.

Tactical Command Basic Organization

Basic organizations for the tactical command are the basic unit operations, basic platoon operations, and breaking contact. Figure 11-19 shows the basic organization.

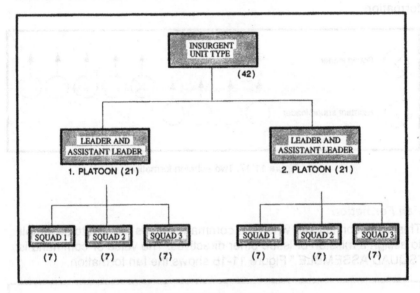

Figure 11-19. Basic organization.

A basic unit operation is when one squad conducts reconnaissance for about an hour before the rest of the unit follows. This reconnaissance squad then occupies key positions for observation to prevent government forces from surprising the unit.

A basic platoon operation is when the first squad provides frontal security, the second squad covers the flanks, and the third assumes the rear guard role.

When breaking contact with the enemy, the first platoon acts as a delaying force until the second platoon withdraws; after which, the first platoon withdraws by squads. Figure 11-20 lists the sounds produced during basic operations. Figure 11-21 shows a typical base camp which is the hub of all operations.

SOUND	DISTANCE THE NORMAL HUMAN EAR CAN HEAR IT
Normal conversation can be heard	90 to 100 meters
Conversation in a low voice	35 to 45 meters
Conversation of words	70 to 80 meters
Normal footsteps	20 to 30 meters
Footsteps over leaves and branches	60 to 80 meters
Sound of coughing	55 to 65 meters
Sound of something being dragged	10 to 20 meters
Cocking a weapon	400 to 500 meters
Rifle shot	2 to 3 kilometers
Automatic weapons fire	3 to 4 kilometers
Machete cutting	150 to 250 meters
Cutting down trees	250 to 350 meters
Trees falling down	750 to 850 meters
Digging trenches	1 to 2 kilometers

NOTE: This chart is based on the approximate distance heard without any outside disruptive influence. at sea level.

Figure 11-20. Sound chart.

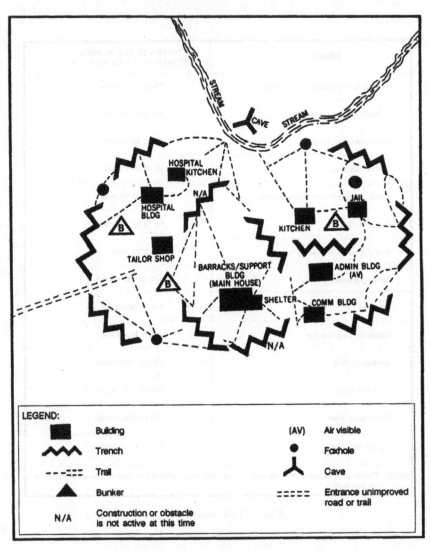

Figure 11-21. Typical base camp.

Intelligence Preparation of the Battlefield Considerations

The doctrine that threat forces use to operate in LIC may not be well known. Your initial IPB effort will probably require you to modify some IPB products to account for the LIC environment. These products can help you plan R&S. Factors not graphically portrayed during the normal IPB process which come into play during LIC include social, political, psychological, and economic factors.

Civilian Population

During LIC operations the civilian population plays a key role. You enhance your R&S plan and the information obtained by gaining the support of the civilian population. You also become familiar with the civilian population's attitude toward their own government and the U.S. forces. In most cases, your R&S asset will observe or monitor groups of civilians to determine if they pose a threat.

The understanding and analysis of the civilian population during the IPB process impacts greatly on the R&S effort. A lesson learned from Operation Just Cause is that "the population cannot read maps nor give grid coordinates." Therefore, when providing the intelligence information part of IPB during prehostilities, build a street map, showing city landmarks; use this with civilian informants.

IPB Products

Products produced during the IPB process impacting on the R&S effort vary depending on the threat. The mission requirement influences the type of overlays and subject categories needed. The following paragraphs cover some LIC IPB products and how they relate to the R&S effort.

Incident Overlay

The incident overlay provides the historical data needed to look for trends and to conduct pattern analysis on the threat. Figure 11-22 shows an incident overlay. You can identify the types of missions the insurgents tend to favor and determine the insurgent AO. This map will show insurgent control or lack of control in specific areas.

Based on this overlay, considerations on where to focus R&S assets are made. Figure 11-23 shows an incident matrix. By constructing an incident matrix, the analyst can determine the times, days, or methods when insurgents will attack targets and can determine their operational trends. The incident overlay will require coordination with the host nation to ensure complete and accurate information.

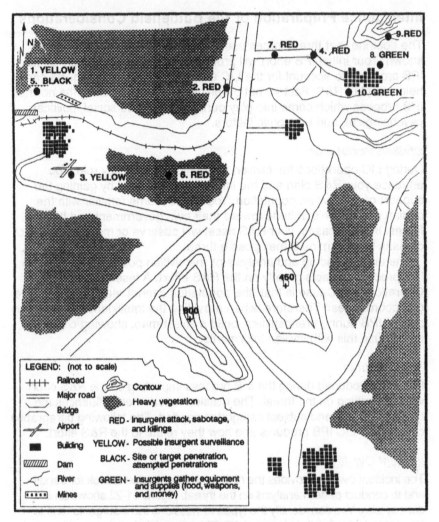

Figure 11-22. Incident overlay.

142

ACTIVITY	DATE/DAY	TIME	LOCATION
1. Surveillance of hydroelectric plant	06 Sep Friday	1000 to 1800	grid * yellow
2. Railroad bridge sabotaged	12 Sep Thursday	2000 to 2300	grid * red
3. Personnel watching airport	08 Oct Monday	0800 to 1500	grid * yellow
4. Bridge sabotaged	21 Sep Friday	2100 to 2200	grid * red
5. Attempted penetration of hydroelectric plant	03 Oct Thursday	0244	grid * black
6. Mine kills two government troops	15 Oct Saturday	1700	grid * red
7. Mine damages bus, 1 killed 5 injured	15 Oct Saturday	1200	grid * red
8. Food stolen from village warehouse	17 Oct Monday	0230	grid * green
9. Mayor executed in village	18 Oct Tuesday	2100	grid * red
10. Police uniforms and weapons stolen	18 Oct Tuesday	0330	grid * green

* Color code activity and plot on map as dot with activity
number next to it. Look for patterns where insurgent forces
are concentrating efforts. Identify insurgent targets and
possible isolation or terrorizing of villages. Analyze matrix
for times or days insurgents may tend to conduct specific
operations.

Figure 11-23. Incident matrix.

Situation Map

The insurgent situation map (SITMAP) is built from the incident overlay. Figure 11-24 shows an LIC SITMAP. The SITMAP adds current
intelligence and activities which indicate insurgent movement, resupply operations, or attacks. You will confirm or deny information on the
SITMAP using R&S assets. These assets:

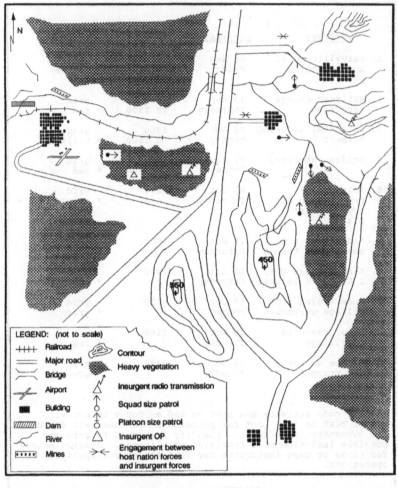

Figure 11-24. LIC SITMAP.

- Monitor insurgent supply routes.

- Monitor radio transmissions.

- Visit civilian communities.

- Patrol LOC.

- Patrol critical sites.

- Provide coordination between local law enforcement and host nation military units.

- Provide sketches of insurgent bases.

When briefing patrols, the SITMAP warns of danger areas such as mines or insurgent controlled areas.

Trap Map

The trap map identifies targets the insurgents will attempt to sabotage or attack. Figure 11-25 shows an LIC trap map. These targets may include:

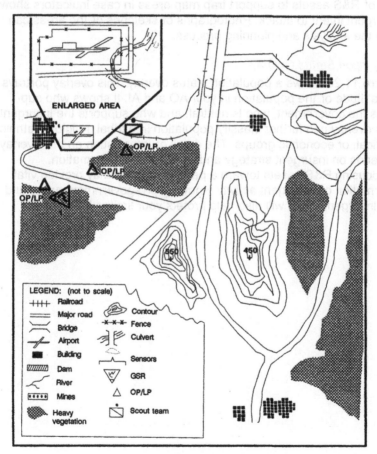

Figure 11-25. LIC trap map.

■ Hydroelectric plants.

■ Weapon storage sites.

■ Airports.

■ Banks.

■ Government offices.

■ Terrain favoring ambushes.

Analyze these areas for insurgent access and escape routes. Preplan use of R&S assets to support trap map areas in case indicators show insurgent intent to attack. Photographs or sketches of the areas can help the analysis and planning process.

Population Status Overlay

Figure 11-26 shows a population status overlay. This overlay portrays the attitude of the population in your AO and AI. It shows who supports the government, who is neutral, and who supports the insurgents. This overlay can further classify population into tribal, religious, ethnic, political, or economic groups. The detail of information on this overlay depends on insurgent strategy and availability of information.

Focusing R&S assets toward a particular group may provide vital information on insurgent activity. Knowing pro-government areas and pro-insurgent areas will also help protect your limited R&S assets.

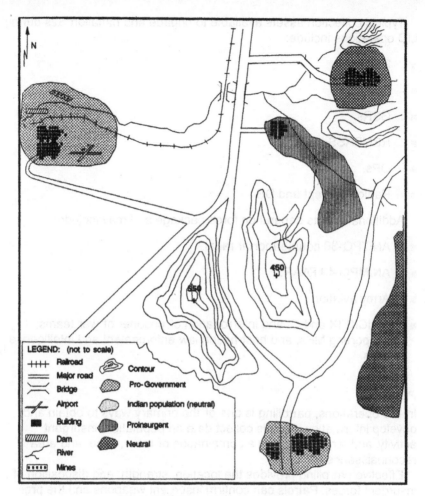

Figure 11-26. Population status overlay.

Assets

R&S assets available during LIC operations depend on mission and
host-nation support. Peacetime contingency operations will require less
R&S assets than counter-insurgency operations. Close coordination
with the host nation will be vital. Information received from local agencies will supplement the R&S plan.

Typical collection assets available to brigade and battalion S2s during LIC operations include:

- Patrols.

- Scouts.

- GSRs.

- REMBASS.

- OPs.

- Radio intercept and DF.

Additional assets depend on force package and may include:

- AN/TPQ-36 countermortar radar.

- AN/MPQ-49 FAAR.

- Army aviation.

- QUICKFIX and CI and interrogation of prisoner of war teams, supporting MPs, and host-nation law enforcement and intelligence units.

Patrols

In LIC operations, patrolling is one of the primary ways to obtain and develop information. Patrols collect data on population, insurgent activity, and terrain by using a combination of route, zone, and area reconnaissance.

Effective patrolling provides the location, strength, and disposition of insurgent forces. Patrols can confirm insurgent weapons and the presence of foreign military advisors. Areas requiring special attention while planning patrols include rivers, streams, and agricultural areas. Patrols conducted around key installations give early warning and prove effective during CR or security missions.

Limitations to consider include communication and security. The primary means of communication for dismounted patrols is the FM battery-powered AN/PRC-77. Also, the patrol's small size leaves it vulnerable to ambush or attack.

The scout platoon gathers information on insurgent forces while conducting patrols or by established OPs in assigned areas. They conduct much the same missions as patrols from the line companies, but scouts usually extend out farther than company patrols. Examples of missions that scouts conduct include:

■ Locating and providing detailed information on insurgent bases.

■ Establishing OPs to monitor these bases while friendly forces move forward to attack.

■ Reporting possible supply routes used by insurgent forces.

Augmenting patrols with attached interrogation assets can add to the patrol's ability to gather intelligence by interrogation or questioning of the local populace.

Augmentation of native scouts familiar with the area provide an advantage. Again, communications and security are primary limitations to scout operations.

GSR

GSR missions in an LIC environment may include continued search of open areas or surveillance of point targets. GSRs are very effective when integrated into R&S plans for installations, bases, and airfield security. They can verify activity detected by other sources (such as OPs, REMBASS, NODS) or vector friendly patrols.

The AN/PPS-5 and AN/PPS-15 are currently found in MI battalions supporting airborne and air assault divisions. The AN/PPS-15 is found in light infantry division MI battalions. System use depends on the AO and the mission.

GSR limitations include LOS to target. This is a key factor when insurgent activity occurs in forested or urban areas.

Extremes in weather such as rain, wind, or snow will degrade their operations.

AN/PPS-5 users must consider the system's weight if the mission is in rough terrain or requires quick movement. It weighs in excess of 110 pounds without batteries, which are 12 pounds each.

Rembass

REMBASS is a valuable asset in the LIC environment. REMBASS gives indications on the amount of traffic along suspected insurgent trails, and it provides early warning when used along routes leading to possible insurgent targets or friendly base camps.

Sensors have a 24-hour, near all-weather capability. The information obtained cues the use of patrols, GSR, or OPs to confirm activity. Considerations when using remote sensors include radio LOS to the relay or monitoring station. The transmitting range is about 15 kilometers for sensors and repeaters, and up to 100 kilometers for airborne repeaters.

Voice Collection Teams

The MI battalion subordinate to heavy, light, air assault, and airborne divisions have voice collection teams capable of supporting the R&S effort. The AN/TRQ-32 TEAMMATE and the AN/PRD-10 provide IEW support. The QUICKFIX and GUARDRAIL (a corps MI brigade asset) can assist in the EW collection effort based on availability.

The AN/TRQ-32(V)(VI)(I) (TEAMMATE) will intercept HF, VHF, and UHF communications. It provides VHF LOB data. The power of the intercepted signal and LOS determine range capability. Limitations to be considered include mobility of the prime mover and security for systems when operating outside a security base.

The AN/PRD-10/11/12 is a mobile radio DF system. It can operate as a single station providing intercept and LOB data. When operating in the net mode with three other stations, the AN/PRD-10 provides intercept and manually computed radio DF fix locations of enemy transmissions. The AN/PRD-10 weighs approximately 80 pounds; its range depends on LOS and the power of the intercepted signal.

When available, the QUICKFIX or GUARDRAIL can conduct airborne DF. These systems have extensive range and can provide locations on enemy transmitters.

Evaluate all available assets within the brigade or battalion capable of supporting the R&S effort. Some assets and their capabilities follow:

- Helicopters resupply, insert, or extract patrols; they also conduct limited route and area reconnaissance.

- The countermortar radar AN/TPQ-36 (DS to the artillery battalion) provides information on insurgent mortar locations.

- Patrols search suspect areas for mortar tubes and ammunition cache sites.

- CI teams provide information on insurgent activities and their intelligence capabilities.

Consider all human sources such as convoy truck drivers, FOs, and personnel from host-nation agencies or units (such as refugee camps, civil affairs checkpoints, local law enforcement, and intelligence.)

Fighting in an LIC environment requires a continuous R&S effort. This effort stresses reporting information to the S2 and disseminating that information no matter how insignificant. It also involves the total force. Figure 11-27 is an example of how these assets are employed in a battalion R&S plan.

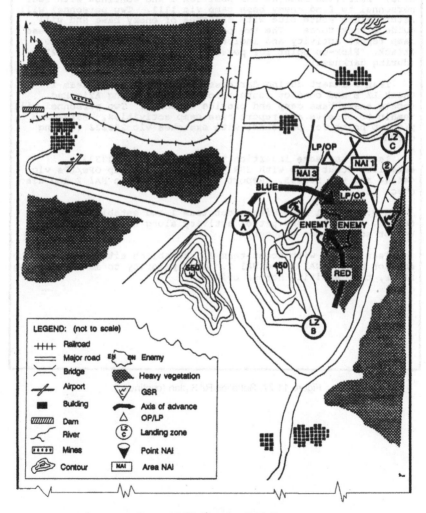

Figure 11-27. Battalion R&S plan.

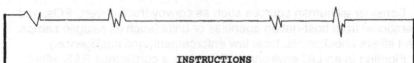

INSTRUCTIONS
FOR BATTALION R&S PLAN
(LIC)

Scout Teams: Team 1 insert during darkness by POV or HMMWV. Check LZ A and conduct route reconnaissance along Axis Blue. Establish base at GSR position 1 and continue with four personnel to find enemy base camp vic 1111. Two personnel will return along Route Blue with sketches of enemy base and will guide attack force. The two personnel remaining at enemy base camp report activity and indications camp may move prior to attack. Pick-up point for personnel with sketches vic 612345 (during darkness).

Team 2 insert during darkness same technique as team 1. Check LZ B and conduct route reconnaissance along Axis Red. Locate enemy base camp and provide sketches. Two personnel will remain to observe and report base camp activities. Establish pick-up point for personnel with sketches vic 222222 (during darkness).

Team 3 use same insertion technique at vic 333333. Establish scout camp with GSR position 2. Set up OPs/LPs vic 444444 and vic 555555. Report enemy movement in NAI 2 and also check LZ C.

GSR: Team 1 move with scout team 1 and establish site vic 666666. Report enemy patrol activity along river vic 777777. Vector scouts upon request.

Team 2 move with scout team 3. Establish site vic 888888 to monitor bridge NAI 3 and NAI 2. Give priority to movement along river, then activity across bridge.

Figure 11-27. Battalion R&S plan (continued).

CHAPTER 12
Electronic Warfare Asset Employment

Normally, division staffs and higher echelons plan for the use of IEW
assets. However, you may be in a situation where you have MI unit
assets either attached or in DS of your unit. In either case, you must be
able to properly direct those assets in support of your R&S plan, as well
as your unit's CR plan. To do that, you should understand:

■   The fundamentals of EW.

■   Who plays what role in EW planning.

■   What IEW assets are needed to help you answer your command-
    er's PIR and IR and support the unit CR plan.

Fundamentals

IEW assets belonging to MI units do three things: they provide combat
information; they provide data which contributes to production of intelligence; and they give your unit an EW capability. EW should be a vital
element of your unit's command, control, and communications counter-
measures (C³CM) program. EW is one way commanders protect their
electronic systems while attacking the enemy's electronic systems. Your
staff should plan for use of EW within three broad mission areas:

■   Defend.

■   Degrade or disrupt.

■   Deceive.

Defend
The defend mission includes your use of electronic counter-
countermeasures (ECCM) to protect your unit's electronic systems.
ECCM includes proper use of signal operation instructions (SOI),
terrain masking, and proper radio and television operator procedures.

FM 24-33 contains detailed information on ECCM. The defend mission also includes ESM to find and target enemy jammers and ECM to screen friendly communications from the enemy.

Degrade or Disrupt

You degrade or disrupt enemy electronic systems by targeting electronic emitters or jamming electronic receivers. Normally, the small number of jammers available to you forces you to be highly selective about which targets to disrupt.

The S3 is in charge of the degrade or disrupt mission. You must support this mission through intelligence and ESM which intercept, identify, and locate potential targets.

Deceive

Electronic deception is normally controlled by division or corps. Deceiving, or electronic deception, provides false information to the enemy through electronic devices. It is intended to induce the enemy into acting against their best interests. Deception is achieved by feeding false or misleading information to enemy electronic sensors, or by transmitting it directly into operational channels. (Normally, this is part of an overall deception plan.) Make sure that what the enemy collects electronically agrees with, or at least does not refute, the overall deception scheme.

IEW systems collect combat information through ESM. Essentially, consider your IEW assets as one more source you can use to help answer your commander's PIR. However, you do not directly task this source, specify which assets do what, nor emplace these assets. You do specify what you want to know, and coordinate with your IEWSE and S3 to make sure your IEW assets do not interfere with your unit's scheme of maneuver. We will address this later in this chapter.

More than likely, your IEW assets will spend significant time supporting or executing the EW degrade or disrupt mission. Look at this mission as consisting of two components: passive and active. ESM is the passive part. That is, your IEW assets work to collect information that will support the ECM, or the active part, of the mission.

ECM consists of jamming and deception. Essentially, jamming delivers a high level of power to an enemy receiver, preventing that receiver from receiving its intended transmission. Your IEW assets must radiate energy to do this. Therefore, they are susceptible to enemy countermeasures.

Deception causes the enemy to misinterpret what is received by electronic systems. Once again, your IEW assets (and any other assets you choose to use) must transmit to deceive. Therefore, deception is active as well. More detailed descriptions of EW are in FM 34-1, Chapter 5.

Planning

Your S3 is in charge of planning and using EW. The S3 is responsible for integrating EW into your unit's scheme of maneuver. However, it is the FSO who must intergrate EW into the overall fire support plan. This becomes especially important for suppression of enemy air defense operations. You must support your S3s EW plan by carefully selecting ESM priorities. (The S2 should assist the S3 with EW planning.) Figure 12-1 is a breakdown of who does what in EW planning.

Finally, the IEWSE officer is the resident expert on your MI unit IEW assets. The IEWSE officer:

- Recommends use.

- Coordinates physical placement.

- Acts as liaison between you and the assets.

- Coordinates EW planning.

- Recommends EW support for fire and maneuver.

Your unit has four electronic options to attack enemy electronic systems:

- Intercept.

- Locate.

- Jam.

- Deceive.

Intercepting provides combat information and technical data on the enemy's electronic systems as well as raw data for processing into intelligence. Technical data supports jamming and electronic deception.

The locating option provides approximate locations of enemy radio and radar antennas. This aids in the use of directional antennas for jamming, and may be used with other information to provide targeting-quality data.

S3	-Friendly situation--electronic deception -Planned operations--ECCM -ECM target priorities -EW HPT identification
FIRE SUPPORT OFFICER	-Fire support integration -EW HPT identification
C-E OFFICER	-MIJI feeder report identification -Taboo, protected frequencies -ECM support to ECCM -Effects on friendly ECCM planning -Electronic deception
S2	-ESM priorities -Enemy situation -Enemy capabilities -HVT identification -EW HPT identification
IEWSE OR MI UNIT	-Electronic OB or EPB - Asset tasking - Asset status - Technical data -Effectiveness assessments

Figure 12-1. EW employment coordination requirements.

Jamming disrupts the receipt or exchange of orders and battlefield information. It can delay the enemy long enough for the commander to exploit a situation that otherwise would have been corrected. Jamming provides a nonlethal alternative or supplement to attack by fire and maneuver. It is well-suited for targets that cannot be located with targeting accuracy, or that only require temporary disruption.

As a general rule, you will want to destroy or jam enemy electronic systems located near your FLOT. Enemy communications systems located farther back will normally be used by enemy planning elements. Therefore, they may be more valuable as a source of information. Figure 12-2 is a list of electronic options listed by enemy echelon and distance from the FLOT.

EW planning follows the normal staff planning process. It begins with the mission and commander's guidance. During the planning process, your staff determines electronic HPTs. You then divide your electronic HPTs into four categories:

- Targets located for destruction (targeting).

- Targets to be jammed.

- Targets to be intercepted for combat information or intelligence.

- Targets to be deceived.

What Type of Intelligence and Electronic Warfare Assets are Needed?

Once you have categorized your electronic HPTs, you need to have a way to tell your IEW assets what you want them to do. Do this by providing your IEW assets a list of priorities on a target list worksheet. Remember to include both ESM (passive) and ECM (active) priorities.

Staff Actions

You and other staff officers determine ESM and ECM priorities by war gaming. Remember that ESM must support ECM. ESM may also help you answer the commander's PIR.

Your S3 ultimately determines ECM priorities (based on staff input). The S2 determines:

- ESM priorities based on your commander's PIR and IR.

COMM NET BY ECHELON	1ST ECHELON						SECOND 2D ECHELON		FRONT
DISTANCE FROM FLOT (km)	0 TO 3	3 TO 6	6 TO 9	9 TO 15	15 TO 20	20 TO 30	30 TO 50	50 TO 100	100 AND UP
COMMAND AND CONTROL	INTCP LOCATE	JAM INTCP LOCATE	JAM INTCP LOCATE	JAM INTCP LOCATE	INTCP LOCATE	INTCP LOCATE	INTCP	INTCP	INTCP
ROCKET, ARTILLERY, AND ASSOCIATED TA	JAM LOCATE	JAM LOCATE	JAM LOCATE	JAM LOCATE	LOCATE	LOCATE	LOCATE	LOCATE	LOCATE
SSM			LOCATE	LOCATE	LOCATE	LOCATE	LOCATE	LOCATE	LOCATE
AIR DEFENSE	JAM LOCATE	JAM LOCATE	JAM LOCATE	JAM LOCATE	LOCATE	LOCATE	LOCATE	LOCATE	LOCATE
INTELLIGENCE	JAM LOCATE	JAM	JAM LOCATE	LOCATE	INTCP		INTCP	INTCP	INTCP
JAMMERS	LOCATE	LOCATE	LOCATE	LOCATE	LOCATE				
ENGINEERS	LOCATE	LOCATE	LOCATE	LOCATE	LOCATE	INTCP	INTCP	INTCP	INTCP
CSS	JAM LOCATE	JAM	JAM	JAM	INTCP	INTCP	INTCP	INTCP	INTCP

NONCOMM RADAR BY ECHELON	1ST ECHELON						2D ECHELON	FRONT
DISTANCE FROM FLOT (km)	0 TO 3	3 TO 6	6 TO 9	9 TO 15	15 TO 20	20 TO 30	30 TO 100	100 AND UP
AIR (SAM AND AAA) DEFENSE	LOCATE	LOCATE	LOCATE	LOCATE	LOCATE	LOCATE	PRIMARY AF RESPONSIBILITY	PRIMARY AF RESPONSIBILITY
WEAPONS LOCATING	LOCATE	LOCATE	LOCATE	LOCATE				
NONCOMM JAMMERS	LOCATE	LOCATE	LOCATE	LOCATE	LOCATE			

NOTE: These options change little during contact in either the attack or the defense.

Figure 12-2. Electronic options.

- The S3s ECM priorities.

- When and where on the battlefield the PIR, IR, and ECM become most important.

The IEWSE officer relays your unit's ESM and ECM priorities to your attached or supporting IEW assets. Figure 12-3 is an example of a completed EW target list work sheet. It shows how you can synchronize those priorities to support the DST. In this example, identifying and locating enemy reconnaissance units of the 141st MRR is the number one ESM priority because the commander's top PIR initially is to locate enemy reconnaissance units.

The second and third ESM priorities are to identify and locate divisional and regimental air defense assets. Note that the first ECM priority is to jam divisional air defense nets. In this case, the third ESM priority supports the first ECM priority. Your EW assets cannot jam those nets without first finding them.

Obviously, the first ECM priority reflects that CAS is important to the success of the mission, and jamming enemy divisional air defense nets supports the planned CAS mission. If you compare the target list work sheet to the DST, you will see the relationship between ESM and ECM priorities and how the unit intends to fight the battle in time.

Electronic Warfare Target List Worksheet
Specifics of the EW target list worksheet follow:

- Time window is the time you want your assets to spend looking for the target; or the start/stop times you want your assets to jam the target.

- Target unit is the specific unit or target you are looking for. The more specific you are, the easier it is for your IEW assets to find it.

- Target location is where you expect the target to be. Location can be based on actual information or on situation templates.

- Target activity is the specific type of communications you want collected or jammed.

- Control mechanism is how you want the target unit to be jammed (such as spot jamming, barrage jamming). Your IEWSE can tell you more about the advantages and disadvantages of each type

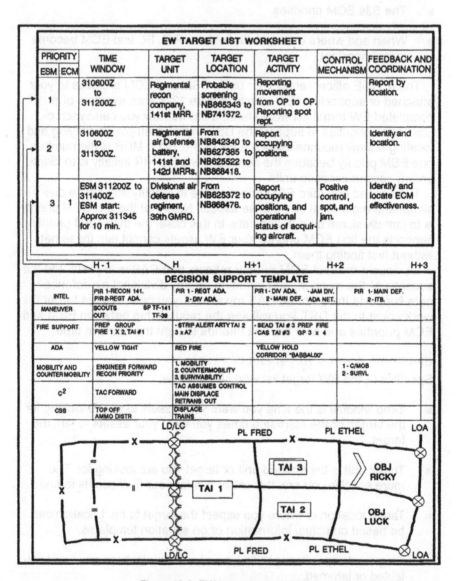

PRIORITY		TIME WINDOW	TARGET UNIT	TARGET LOCATION	TARGET ACTIVITY	CONTROL MECHANISM	FEEDBACK AND COORDINATION
ESM	ECM						
1		310600Z to 311200Z.	Regimental recon company, 141st MRR.	Probable screening NB865343 to NB741372.	Reporting movement from OP to OP. Reporting spot rept.		Report by location.
2		310600Z to 311300Z.	Regimental air Defense battery, 141st and 142d MRRs.	From NB842340 to NB627385 to NB625522 to NB868418.	Report occupying positions.		Identify and location.
3	1	ESM 311200Z to 311400Z. ESM start: Approx 311345 for 10 min.	Divisional air defense regiment, 39th GMRD.	From NB625372 to NB868478.	Report occupying positions, and operational status of acquiring aircraft.	Positive control, spot, and jam.	Identify and locate ECM effectiveness.

EW TARGET LIST WORKSHEET

H - 1   H   H+1   H+2   H+3

DECISION SUPPORT TEMPLATE

INTEL	PIR 1-RECON 141. PIR 2-REGT ADA.	PIR 1 - REGT ADA. 2 - DIV ADA.	PIR 1 - DIV ADA. - JAM DIV. 2 - MAIN DEF. ADA NET.	PIR 1- MAIN DEF. 2 - ITB.
MANEUVER	SCOUTS SP TF-141 OUT TF-39			
FIRE SUPPORT	PREP GROUP FIRE 1 X 2, TAI #1	- STRIP ALERT ARTY TAI 2 3 x A7	- SEAD TAI # 3 PREP FIRE - CAS TAI #3 GP 3 x 4	
ADA	YELLOW TIGHT	RED FIRE	YELLOW HOLD CORRIDOR "BABBALOO"	
MOBILITY AND COUNTERMOBILITY	ENGINEER FORWARD RECON PRIORITY	1. MOBILITY 2. COUNTERMOBILITY 3. SURVIVABILITY		1 - C/MOB 2 - SURVL
C²	TAC FORWARD	TAC ASSUMES CONTROL MAIN DISPLACE RETRANS OUT		
CSS	TOP OFF AMMO DISTR	DISPLACE TRAINS		

Figure 12-3. EW target list worksheet.

of jamming and can recommend which kind will best support your mission.

■ Feedback and coordination is exactly what you are looking for; for example, location and identification of the target and effectiveness of jamming. Check with your S3 to see what kind of feedback is required. Again, your IEWSE can help in this area.

Target List Worksheet: Reconnaissance and Surveillance Plan

An EW target list worksheet is the equivalent of an R&S plan for your IEW assets. It tells your assets what they should look for, when they should look for it, and what and when to jam. The EW target list worksheet should be a total staff effort among you, the S3, the FSO, and the IEWSE.

Remember, you should not be concerned with the details of how to collect the information, such as which specific asset should do what, and where each asset should go. Those details are left to the attached or supporting MI unit and your IEWSE. Simply tell them what you want, and let them figure out how best to do it.

You must, however, make certain that the emplacement of your IEW assets does not interfere with your unit's mission. Therefore, make sure your IEWSE coordinates all IEW positions with your S3. Also, remember to continually monitor those IEW positions so that they do not get over-run or outdistanced by maneuver units.

Keep your IEWSE updated on the enemy frontline trace and insist on frequent status reports on your IEW assets. Figure 12-4 is a list of organic or supporting MI units by echelon. Figure 12-5 is an electronic attack options chart. Figure 12-6 shows IEWSE officer responsibilities. FM 34-40 provides a detailed discussion of EW operations.

ELECTRONIC WARFARE UNITS			
ECHELON	MI ORGANIZATION	MAJOR EW UNIT	EW SUBUNIT
CORPS	BRIGADE	MI BN (TE) MI BN (TE) (RC) MI BN (AE)	EW CO EW CO (COLL) AND EW CO (ECM) AVN CO (EW)
HEAVY DIVISION	BATTALION	C&J CO EW CO	C&J PLATOON SIGINT PROCESSING PLT QUICKFIX FLT PLT (OPCON)
LIGHT DIVISION	BATTALION	COLL CO	VOICE COLL PLT QUICKFIX FLT PLT (OPCON)
AIR ASSAULT DIVISION	BATTALION	HHOC C&J CO	QUICKFIX FLT PLT C&J PLT AND NONCOMM PLT
AIRBORNE DIVISION	BATTALION	C&J CO	C&J PLT AND NONCOMM PLT QUICKFIX FLT PLT (OPCON)
ACR	COMPANY		C&J PLT QUICKFIX FLT PLT (OPCON)
SEPARATE BRIGADE	COMPANY		COLL PLT (VOICE) VHF ECM PLT

Figure 12-4. Supporting IEW units by echelon.

LOCATION NAI AND TAI	INDICATOR ACTIVITY	PROB TIME	ENEMY			ELECTRONIC ATTACK					FRIENDLY EW SYSTEM	SYSTEM SITE LOCATION	C3 SITE LOCATION	REMARKS
			TYPE NET	TYPE EMITTER	FREQUENCY	DF OR LOCATE	JAM	MONITOR ID	DECEIVE					
PIR	Movers Shooters Sitters Emitters		Recon C2 Manuever Arty/AD Engineer REC/EW Reserves	Pre-planned On call Target of oppor-tunity									1. PIR (number below) 2. Coordination 3. Deconfliction 4. Reports 5. Communication channels and means 6. (Interoperability) 7. (Enemy code words, net struc-tures, crypto systems, and radars, associated with deployments and units) 8. Jamming effec-tiveness

PIR: Precise tasking, directed toward specific emitters with carefully prescribed reporting procedures (FM 34-10A, p B-24).

Figure 12-5. Eletronic attack options.

163

1. Integrates IEW missions with brigade operations.

    a. Ensures collection and jamming missions support the commander's intent, concept of operations, scheme of maneuver, and fire support plan.

        (1) LOS analysis for initial deployment locations, specifically:

            (a) High ground for C&J.

            (b) LOS from jammers and collectors to target (NAI).

            (c) LOS between collectors for netting.

            (d) LOS from jammers and collectors to C3 sites for reporting.

        (2) LOS analysis for subsequent redeployment locations in accordance with overall tactical scheme of maneuver. Same as above.

    b. IEWSE officer studies brigade OPORD with special attention to items above, especially unit boundaries and phase lines.

    c. Coordinates with brigade S2s to identify specific PIR, IR, and NAI to confirm or deny enemy COAs, with special attention to—

        (1) How the S2 estimates enemy COAs to develop.

        (2) Templated enemy forces and IPB.

        (3) Probable indicators and enemy forces expected to deploy at each NAI.

        (4) Associated enemy emitters and electronic clusters at each NAI.

        (5) Probable time line(s).

        (6) Type nets and frequencies (technical control and analysis element).

Figure 12-6. Brigade IEWSE officer responsibilities.

(7) Location of specific collection assets to cover specific NAI above.

(8) Impact of weather on operations.

d. Coordinates with S2 to ensure collection operations reflect commander's specific PIR and IR; and are synchronized with the overall R&S plan, with special attention to--

(1) Unit scouts, patrols, OPs/LPs.

(2) GSRs.

(3) REMBASS.

(4) Division long-range surveillance units.

(5) Cavalry operations.

(6) USAF reconnaissance.

(7) Interrogation of prisoners of war and CI.

(8) Fire support.

e. Coordinates with brigade S2 and brigade communications-electronics staff officer (CESO) to ensure current brigade challenge and password is passed to MI battalion S3.

f. Coordinates with brigade S3 to ensure IEW operations are synchronized with brigade scheme of maneuver, with special attention to--

(1) Locating jamming assets to serve as an effective combat multiplier as part of overall scheme of maneuver and fire support plan.

(2) Coordinating operational sites and fallback positions with maneuver elements.

(3) Coordinating times and routes for deployment or redeployment.

Figure 12-6. Brigade IEWSE officer responsibilities (continued).

(4) Planning redeployment of MI ground C&J assets by bounding as part of overall scheme of maneuver.

(5) Planning integration of airborne QUICKFIX to cover times and locations not covered by ground assets.

g. Coordinates with brigade S4 for logistic support, with special attention to--

(1) Attachment of MI assets, as appropriate.

(2) Roll up of all MI battalion assets operating in brigade ACR, from MI battalion S3 and S4.

h. Coordinates with brigade CESO for--

(1) Proper CRYPTO fill and frequency for brigade operations and intelligence (O&I) net.

(2) Protected frequencies to prevent electronic fratricide.

(3) Deconfliction of IEW operational and C3 sites with signal units.

(4) Location of jammers.

(5) Relays.

i. Coordinates with brigade FSO for--

(1) Deconfliction of IEW operational and C3 sites with fire support units.

(2) Fire support planning to avoid engagement areas.

(3) Family of scatterable mines (FASCAM) targets.

(4) Use of tactical fire direction computer system (TACFIRE) net to pass sanitized reports quickly.

(5) Fire support radar locations.

(6) Protective fires to cover withdrawal of IEW assets, as required.

Figure 12-6. Brigade IEWSE officer responsibilities (continued).

j. Coordinates with brigade engineer for--

(1) Location of current or planned mine fields and obstacles.

(2) Brigade classifications.

(3) Route reconnaissance.

(4) River crossing sites and classification.

k. Coordinates with brigade ADA officer for--

(1) Deconfliction of IEW operational and C3 sites with ADA units.

(2) FAAR locations.

(3) QUICKFIX air corridors, loiter times, and locations.

(4) Identification, friend or foe (IFF) (radar).

(5) Current ADA weapons control status and location.

l. Coordinates with brigade NBC officer for--

(1) Current contaminated areas.

(2) Planned NBC targets.

(3) Smoke and obscurant employment.

m. Coordinates with brigade MP officer for current condition and status of routes, traffic control measures, and support to the reconnaissance, intelligence, surveillance, and target acquisition (RISTA) effort.

n. Coordinates with brigade ALO for--

(1) Timely in-flight reports.

(2) Latest weather updates.

(3) Current and planned air and ground engagement areas.

Figure 12-6. Brigade IEWSE officer responsibilities (continued).

2. Assists the brigade S2 in formulating the EW annex of the brigade OPLAN and OPORD.

3. Serves as a critical liaison officer between MI battalion and the maneuver brigade.

a. Keeps the MI battalion S3 informed of above data from brigade staff.

b. Serves as functioning member of brigade orders group.

c. Provides copies of maps and overlays from brigade orders group.

d. Monitors brigade O&I net and brigade command net for latest developments.

e. Notifies MI battalion S3 in a timely manner of all brigade developments and redeployments.

4. Keeps brigade commander and staff informed of latest intelligence depicting the enemy situation:

a. Monitors tasking and reporting net for tactical reports.

b. Sanitizes tactical reports and passes combat information to brigade commander and staff in 15 minutes or less from time of receipt.

Figure 12-6. Brigade IEWSE officer responsibilities (continued).

APPENDIX A

Management Tools for Reconnaissance and Surveillance Operations

This appendix contains the following checklists, formats, and reports used in planning and supervising R&S missions.
NOTE: This appendix can be reproduced and used in the field as a pocket guide.

- Figure A-1. Example of an intelligence estimate in matrix format.

- Figure A-2. R&S tasking matrix.

- Figure A-3. Different versions of the R&S tasking matrix.

- Figure A-4. R&S checklist.

NOTE: The purpose of the R&S checklist is to make sure complete coordination is conducted for all R&S operations.

- Figure A-5. Collection plan format.

- Figure A-6. IEW Asset redeployment matrix.

- Figure A-7. Reconnaissance asset utilization matrix.

- Figure A-8. A patrol plan.

- Figure A-9. Patrol report format.

- Figure A-10. Hints on debriefing patrols.

- Figure A-11. GSR or REMBASS plan format.

- Figure A-12. GSR or REMBASS briefing checklist.

- Figure A-13. Standard collection asset request format (SCARF) basic format.

- Figure A-14. Mission report format.

AS OF CURRENT INTELLIGENCE ESTIMATE

1. Weather: For period from _____ to _____.

BMNT from _____ to _____.   EENT from _____ to _____.

Sunrise _____.   Sunset _____.   Temp from _____ to _____.

Winds: Speed from _____ to _____ knots. Direction_____.

Night operations: Skies cloudy clear. % of moon illum _____.

Trafficability: Poor good excellent.

Precipitation: Snow rain sleet hail.

Maximum precipitation per month _____ inches.

Average precipitation per month _____ inches.

2. Terrain: From_____ to _____.

Observation and fields of fire:

Concealment and cover:

Obstacles:

Key terrain

    Decisive terrain:

    Other key terrain:

Avenues of approach:

Other:

Figure A-1. Example of intelligence estimate in matrix format.

```
3.   Enemy in brigade sector

        Map sheets:

        Intelligence overlay:


    Committed forces:



    Reinforcements:




    Recent activities:


    Strengths:


    Weaknesses:


    Enemy frontline trace:


    Second line of defense:


    Probable courses of action:


    Enemy NBC:          Possible        Probable        Imminent
                                        Next _____ hours.
```

Figure A-1. Example of intelligence estimate in matrix format (continued).

PRIORITY (matches PIR no)	NAI (grid)	START STOP	SIR	A	B	C	D	E and AT				COORDINATION	REPORTS

DTG: _____

(CLASSIFICATION)

MISSION: _____

(CLASSIFICATION)

NOTE: A maneuver battalion S2 or S3 tasking subordinate units would change SIR to SOR. S2s use this matrix to coordinate and manage the R&S effort.

Figure A-2. R&S tasking matrix.

BRIGADE R&S TASKING MATRIX

UNIT TASKING	PRIORITY	NAI	LOCATION	REPORTING REQUIREMENT EVENT OR INDICATOR		REMARKS

BRIGADE OR BATTALION R&S TASKING MATRIX

PRIORITY	NAI	LOCATION	REPORTING REQUIREMENT EVENT OR INDICATOR	COLLECTION UNIT OR ASSETS									REMARKS

Figure A-3. Different versions of the R&S tasking matrix.

I. Planning Process.

   A. Initial Requirements:

      1. Did higher headquarters provide tasking requirements?

      2. Was the commander's PIR and IR stated and included?

      3. Did the commander provide R&S intent?

      4. Did the S2 brief the staff on enemy collection capabilities?

      5. Were other staff tasks performed?

   B. SIR Developed (IPB Driven):

      1. Did S2 identify air or ground AAs?

      2. Do situation and event templates reflect probable or prioritized enemy COAs?

      3. Was NAI developed in detail? (What is expected? When? Where?)

      4. Were collectable indicators at NAI developed?

      5. Were SIR developed from NAI and indicators?

      6. Were reporting requirements developed for priority collection missions to allow the commander time to change plan?

   C. Possible Collectors Analyzed:

      1. Did S2 coordinate with staff, S2, and G2 to identify all available collection assets?

      2. Did S2 analyze asset capabilities to develop collection requirements based on range to target, time available, target characteristics, terrain, weather, enemy (obscurants use), and communication?

      3. Did S2 analyze collection redundancy (is it necessary)?

Figure A-4. R&S checklist.

          4. Did staff identify support requirements
(communication nets, retransmission, fire support, logistic
support, special equipment support)?

          5. Did S2 identify gaps in collection?

          6. Did S2 backbrief S3 or the commander on R&S
concept?

          7. Were warning orders sent to appropriate assets?

          8. What were timelines?

               a. When was mission received?

               b. What is NLT for execution?

               c. When was templating done?

               d. When was tentative plan made? Backbriefed?

               e. When were warning orders issued?

               f. When was initial reporting needed?

               g. Who was in charge of R&S planning?

               h. Who was in charge of CR planning?

II. Preparation for R&S Operations.

     A. Specific Collection Instructions:

          1. What assets were available? Used?

               a. Scouts.

               b. GSR.

               c. Patrols.

               d. OPs/LPs.

               e. FOs.

               f. Infantry.

Figure A-4. R&S checklist (continued).

```
        g.   Armor.

        h.   AT.

        i.   Aviation.

        j.   Engineer.

        k.   Signal.

        l.   Cavalry.

        m.   EW.

        n.   MP.

        o.   Other.

    2.   Did the S2 provide detailed instructions to tasked
assets?  Did the instructions include--

        a.   Who is tasked?

        b.   What to look for?

        c.   Why to look?

        d.   When to look?

        e.   Where to look?

        f.   What you could expect to see?

        g.   How to get there?

        h.   Who to coordinate with?

        i.   Passage of lines and recognition signals?

        j.   Reporting requirements?

        k.   Friendly assets in AO?

        l.   Resupply?

        m.   Obstacles?
```

Figure A-4. R&S checklist (continued).

n. MEDEVAC?

o. NBC hazards?

3. Was the collection location appropriate (concealment, collectable)?

4. Were there sufficient control measures included to control assets during mission?

5. Did the S2 request assistance from higher headquarters for identified collection gaps?

6. Did the R&S plan cover all collection requirements?

7. Were assets overtasked?

8. Was redundancy appropriate for this mission?

B. Coordination:

1. What is format of plan (collection plan, overlay, matrix)?

2. Were direct or indirect fires or jamming coordinated between staff and S2?

3. Was a CI vulnerability assessment made?

4. Was additional equipment (special) planned for?

5. Were communication nets established to meet reporting needs?

6. Were commanders and staff briefed on plan before execution?

7. Did commander or S3 approve final plan?

8. Did assets know specific requirements (PIR and IR)?

9. Was plan disseminated to all involved or those with a need to know (FRAGO, overlay)?

10. Was plan sent to higher headquarters?

Figure A-4. R&S checklist (continued).

   C.   Asset Internal Coordination:

        1.   Was equipment checked?

        2.   Were internal procedures clarified?

        3.   Did coordination between assets occur?

        4.   Was mission rehearsed?

        5.   Was plan developed far enough in advance for assets to prepare or rehearse?

        6.   Was plan developed in time for higher headquarters to review?

 III.   Execution.

   A.   Continuity of R&S and CR Operations:

        1.   Did unit plan provide for operations when scout or other R&S assets are inoperable?

        2.   Did unit SOP provide for operations during briefings, debriefings, or rehearsals?

        3.   Are units and leaders cross-trained to facilitate substitutions or replacement of scouts?

   B.   Asset and Unit Response:

        1.   Did assets depart and set up on time?

        2.   Did assets use concealment, cover, and camouflage?

        3.   Were assets able to observe enemy undetected?

        4.   Was low-level deception used?

        5.   What were meteorological report requirements?

        6.   Were enemy locations pinpointed?

        7.   Was objective reconnoitered?

Figure A-4. R&S checklist (continued).

178

8.   Were obstacles identified and marked?

9.   Were routes marked?

10. Was enemy reconnaissance located?

11. Were CR missions performed?

12. Did assets help with C2 during attack?

13. Did assets help direct or control fires?

14. Was terrain reconnoitered?   (Trafficability reported?)

   C. Reporting:

1.   Were reports timely, accurate, and concise?

2.   Were assets debriefed?

   D. Results:

1.   Did S2 plot asset reports (track results of plan)?

2.   Did S2 identify inadequately tasked or unproductive assets and change tasking (with approval of commander or S3)?

3.   Did reports or analysis answer PIR and IR?

4.   Was R&S plan updated and recoordinated?

5.   Were templates updated?

   E. Dissemination:

1.   Was commander briefed on answer to PIR?

2.   Has commander updated PIR?

3.   Did units get intelligence based on priority?

4.   Did higher headquarters get answers to taskings?

5.   Did assets receive feedback on level of success?

Figure A-4. R&S checklist (continued).

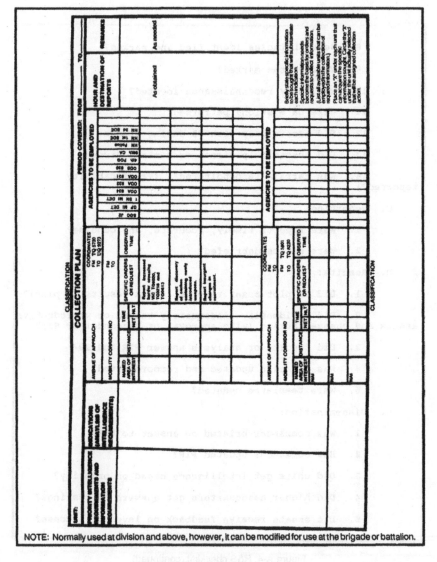

NOTE: Normally used at division and above, however, it can be modified for use at the brigade or battalion.

Figure A-5. Collection plan format.

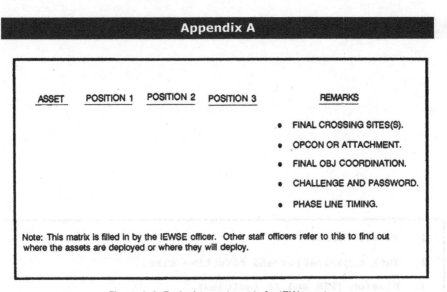

Figure A-6. Redeployment matrix for IEW assets.

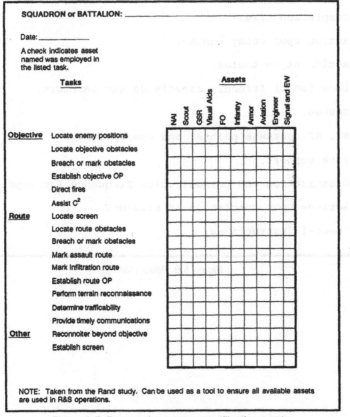

Figure A-7. Reconnaissance asset utilization matrix.

1. Patrol number.

2. Task organization and effective time.

3. Mission (PIR and IR included).

4. Start time.

5. Completion time.

6. Action upon enemy contact.

7. Action at obstacles.

8. Location of friendly minefields and barriers.

9. Routes.

10. SP, RP, passage points, and checkpoints.

11. Fire support.

12. Organization and communication frequency for reporting.

13. Actions upon completion of mission.

14. Special instructions.

Figure A-8. Patrol plan.

Patrol reports are prepared by the S2 section based on information reported by the patrol during debriefing. All pertinent information is included in the report to ensure completeness. The report is then disseminated in accordance with appropriate SOP.

(DESIGNATION OF PATROL)
TO: _____
MAPS: _____

A. Size and composition of patrol.
B. Mission.
C. Time of departure.
D. Time of return.
E. Routes out and back.
F. Terrain: (Description of the terrain: dry, swampy, jungle, thickly wooded, high brush, rocky; deepness of ravines and draws; condition of bridges as to type, size, and strength; effect on armor and wheeled vehicles.)
G. Enemy: (Strength, disposition, condition of defense, equipment, weapons, attitude, morale, exact location, movements, and any shift in disposition; time activity was observed; coordinates where activity occurred.)
H. Any map corrections.
I. (Not used.)
J. Miscellaneous information (including aspects of NBC warfare).
K. Results of encounters with enemy: (Enemy prisoners and disposition, identifications, enemy casualties, captured documents, and equipment.)
L. Condition of patrol (including disposition of any dead or wounded).
M. Conclusions and recommendations (include to what extent the task was accomplished and recommendations as to patrol equipment and tactics).

_____ _____ _____
Signature   Rank   Unit of Patrol Leader

N. Additional remarks by interrogator.

_____ _____ _____ _____
Signature   Rank   Unit of Interrogator   Time

O. Distribution.

Figure A-9. Patrol report format.

1. Make the subject comfortable.

2. Debrief orally.

3. Debrief as soon as possible.

4. Do not ask leading questions.

5. Make sure PIR and IR are given to the patrol before it goes out.

6. Was intelligence accurate? If not, what were the inaccuracies?

7. Was the map accurate? If not, what were the inaccuracies?

8. If aerial photos were used, was the imagery analysis accurate?

9. What was the condition of trails? Did the trails show signs of recent use?

10. Were rivers or streams crossed or followed?

11. What were stream and river conditions.

12. Give location of the following and were they in recent use?

    o Bridges, by type.

    o Fords.

    o Roads, trails.

13. Local contacts:

    o Name of village, city, or town.

    o Number, location.

    o Were they friendly? Indifferent? Hostile?

    o Was there any previous government contact?

    o Have they moved recently? If so, why?

    o Did they give any information?

Figure A-10. Hints on debriefing patrols.

14. Enemy contacts:

    o Where and when.

    o Number, race, and sex?

    o How were they dressed?

    o Were they carrying packs? How were they armed?

    o Activity? If moving, in what direction?

    o Any equipment or documents recovered?

    o Casualties (own and enemy)?

    o Enemy casualties identified? Disposition of bodies?

    o Any prisoners or crossovers?

    o Any observed (detected) ground or airborne R&S or other intelligence collection activity?

15. Camps:

    o Where, when, and how was camp discovered?

    o How sited? How many huts or buildings?

    o What type? How long ago were they built?

    o In use? Food dumps?

    o Defenses constructed? Approach and escape routes?

    o Weapons, ammunition, or tools?

    o Signs of radios having been used?

    o Electricity? Generators? Used batteries?

    o Documents?

    o Printing press? Copiers? Used carbon paper?

    o What was done with the camp?

Figure A-10. Hints on debriefing patrols (continued).

16. Cultivated areas:

   o Time, date, location of discovery?

   o Size and shape?

   o Was camouflage attempted?

   o What crops?

   o How old? Signs of harvesting?

   o When last tended?

   o Signs of habitation? Tools? Machinery?

   o Trails in area?

   o What was done to the cultivation?

Figure A-10. Hints on debriefing patrols (continued).

1. Mission.

2. Time required to be operational.

3. Routes to GSR site or REMBASS emplacement areas.

4. Location of primary or alternate and subsequent GSR sites and REMBASS strings and fields.

5. Left and right scan limits of GSRs in miles; point target location.

6. Withdrawal routes to subsequent GSR sites.

7. Location of friendly minefields and barriers.

8. Action upon enemy contact.

9. Passage points and checkpoints.

10. Fire support planning.

11. Organization and communication frequency for reporting.

12. Special instructions.

Figure A-11. GSR or REMBASS plan format.

---

1. Friendly and enemy situation.

2. Mission.

3. Maps, photos, overlays.

4. Expected targets (type).

5. Required reports.

6. Reporting procedures.

7. Communications.

8. Security.

9. Food service.

10. Maintenance support.

11. Sensor emplacement sites.

12. Monitoring sites.

13. Mission security implant.

14. Proposed radar site locations.

15. Surveillance sectors.

16. Operation periods and schedules.

17. MEDEVAC.

18. Passage of lines.

Figure A-12. GSR or REMBASS briefing checklist.

Standard Collection Asset Request Format

Asset managers use the standard collection asset request format (SCARF) for intelligence requirements tasking and for requesting information from higher or adjacent headquarters. At corps and division, intelligence requirements tasking is directed toward MI commanders and commanders of other elements of the combat force capable of collecting the information.

Intelligence requirements tasking provides the selected unit with a specific requirement, but not with specific instructions for carrying out the mission. The SCARF tells you:

■ What information is needed.

■ Where to look for the movers, shooters, sitters, and emitters.

- When to look.

- When the information is needed.

- What to look with (in some cases).

A. Requester number.

B. Originator priority.

C. Activity or target type (area emitter and size [point, areas, or unit]).

D. BE number, ELINT notation or case.

E. Location (if known or last known).

F. Duration:

o Start date-time.

o Frequency.

o Stop date-time.

o Latest acceptable date-time for information utility.

G. Location accuracy:

o Required.

o Acceptable.

H. PIR and IR desired.

I. Justification.

J. Remarks (include disciplines and collectors recommended).

Figure A-13. SCARF basic format.

Joint Tactical Air Reconnaissance and Surveillance Request Form (DD Form 1975)

This form is used to request a joint tactical air reconnaissance or surveillance mission. The form states requirements, identifies needs, and, occasionally, specifies the actual air asset to perform the mission.

Any echelon initiates the request. It is used for both immediate and preplanned mission requirements. The requestor completes section of the request for each specific mission. Normally, these requests are transmitted by electrical means. The headquarters receiving the request adds information required to validate or complete the request. Obtain DD Form 1975 through your S1.

Mission Report

Air units include strike or attack, reconnaissance or surveillance, airlift, observation, and helicopter. Air units use the mission report to report the results of all missions and significant sightings along the route of the flight.

The mission report amplifies the inflight report and is normally submitted within 30 minutes after aircraft landing to:

- The tasking agency.

- The requesting unit or agency.

- Other interested organizations.

When the EW air task or mission is completed, intelligence personnel conduct a briefing and submit a mission report.

HEADING

PRECEDENCE

ORIGINATING AGENCY

ACTION ADDRESSEES

INFORMATION ADDRESSEES

SECURITY CLASSIFICATION/CODE

MISSION REPORT (Number) DATE-TIME GROUP

BODY

1. **AIR TASK/MISSION NUMBER OR NICKNAME**. Reference the request number, FRAGO number, or directive causing initiation of the mission.

2. **LOCATION IDENTIFIER**. Target number, line number, approved target designator/identifier, or coordinates of the target or sighting being reported.

3. **TIME ON TARGET/TIME OF SIGHTING**. Report all times by date-time group, using ZULU time unless otherwise directed.

4. **RESULTS/SIGHTING INFORMATION**. This item should contain the pilot and aircrew evaluation of expected results (for example, percent destroyed, number and type destroyed, or percent of coverage). It should contain concise narrative information on significant sightings (for example, unusual or new enemy equipment or concentrations of enemy forces observed to include number, speed, and direction [if applicable]).

5. **REMARKS**. Includes information not specifically mentioned in above items (for example, enemy defenses encountered; weather data; hostile meaconing, intrusion, jamming, and interference attempts).

Figure A-14. Mission report format.

Appendix B

Example of the Reconnaissance and Surveillance Process

The objective of R&S planning is the collection of information the commander needs in order to fight and win the battle. Planning results in the coordinated efforts of all intelligence resources integrated into one collection effort.

Planning Process

The planning process includes:

- Determining requirements.

- Assigning priorities.

- Allocating the resources to satisfy each requirement.

This appendix will assist commanders and staffs in understanding the process used to develop, implement, and execute an R&S operation.

The process described in this appendix is a deliberate one which can be used when sufficient time is available. In a hasty planning process, the procedures can be modified. Most of the products described here will not be done formally, but the steps involved should still be applied mentally.

Maneuver Brigade Scenario

The following is the scenario for a maneuver brigade in a high-intensity conflict.

COL Link Gayagas, Commander, 1st Brigade, 52d Infantry Division (Mech), had just received the divisions OPORD. Based on the division commander's concept of operations and intent, COL Gayagas knew his brigade was in for a hard time in accomplishing the mission. The brigadets mission is to conduct a supporting attack in the southern zone of the division's AO.

The brigade is to seize defensible terrain. This will allow the division to prepare for a defense and destroy a reinforcing combined arms Army soon to be committed. To support the division's main attack, the division commander wants the brigade to draw the commitment of the

41st guards motorized rifle division's (GMRD) reserve, the 35th tank regiment (TR), into the brigadets zone. The 35th TR is the only threat to any major drive by the 52d Infantry Division to the north.

While still at the division OPORD briefing, COL Gayagas took advantage of a short break to instruct the brigade S3, MAJ Booth, to call the brigade TOC and give them a warning order for the upcoming mission. MAJ Booth provided the brigade TOC with the type mission, boundaries, and the brigade's objective, as assigned by higher headquarters. This was to allow the staff to begin work on the mission, particularly the S2 who needed the additional time to develop the intelligence products to support the brigade's IPB process.

MAJ Baker, the brigade's S2, immediately began to orchestrate the intelligence system to support the upcoming mission. He directed SGT Hockings, the section intelligence analyst, to develop an MCOO of the AO, and, for initial planning purpose, to include in the MCOO the analysis of the AI extending 5 kilometers to the flanks and 10 kilometers forward of the AO. MAJ Baker also directed his assistant, CPT Roberts, to call the G2 shop or division operations and intelligence (O&I) and get as much information as possible on the enemy situation.

By the time the brigade commander returned from the division OPORD meeting, MAJ Baker had developed a good idea of the enemy situation; and since he understood the informational requirements associated with the type mission assigned, he was ready to support the brigadets decision-making process.

Brigade Planning

The brigadets planning staff was assembled quickly upon the return of the commander. COL Gayagas provided all the information he had that was not published in the OPORD. He also provided the higher commander's intent and guidance, insights, and concerns, along with some available options.

COL Gayagas was particularly concerned about the brigade's ability to create a situation which would force the enemy motorized rifle divisions (MRD) commander to commit his reserve into his brigade sector. He knew he had to find an enemy weakness and exploit it quickly; to do that, he needed detailed information on the enemy disposition. COL Gayagas gave MAJ Baker the PIR:

1. What is the 15th guards motorized rifle regiment (GMRR) defensive disposition?

2. Is there a weakness in the 15th GMRR defensive disposition? If so, where?

3. Where are the artillery battalions comprising the 15th GMRR regimental artillery group?

4. Will the 41st GMRD commander direct any of his gunships against 1st Brigade? If so, when?

5. Where and when will the 35th TR be committed?

6. Will the enemy employ chemical munitions against 1st Brigade? If so, when and where?

COL Gayagas provided his planning guidance, stressing his PIR. He left his second in command, LTC Larcom, to initiate the planning process and coordinate those staff actions requiring operating system integration. LTC Larcom provided each staff element with its corresponding portion of the division OPORD.

He informed the staff they had two hours to go through the mission analysis process in their respective area of responsibility and to be back at the end of those two hours to review the results of their analysis.

S2 Analysis Process

With the commander's PIR in hand, MAJ Baker began to develop the products needed to support the accomplishment of the mission. When MAJ Baker arrived at his 577, CPT Roberts and SGT Hockings were refining the initial situational template. They were comparing their product to the OB holdings on the enemy unit facing 1st Brigade. MAJ Baker informed CPT Roberts he had received the division's intelligence products and commander's PIR and needed him to assist in the mission analysis process.

They both understood the higher commander's intent and knew the informational requirements associated with the offensive operation being conducted by the brigade. They completed the mission analysis process as it pertained to the intelligence system. MAJ Baker was about to leave to meet with the orders group to present the result of his analysis. He took with him the MCOO and the enemy situation template. Figure B-1 shows an MRR situation template (based on a prepared defense). He instructed CPT Roberts to start developing the brigade R&S plan.

CPT Roberts began his efforts by analyzing the commander's PIR. Using the enemy situation template, CPT Roberts took the PIR and began to associate them with indicators of enemy COAs. At the same time, he identified those PIR which could be satisfied with organic, assigned, or attached collection assets, and those PIR for which he

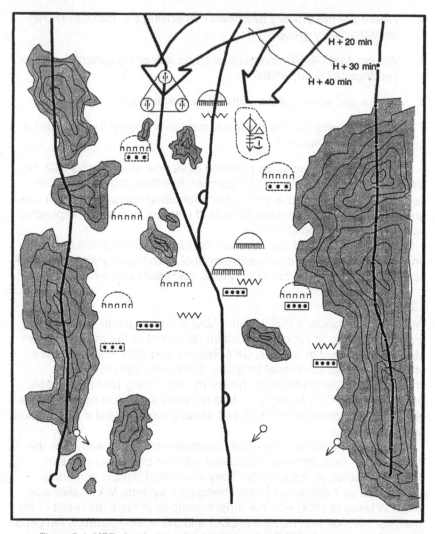

Figure B-1. MRR situation template based on a prepared defense.

would have to submit an RII to higher headquarters. The PIR analysis conducted by CPT Roberts reflected:

PIR: What is the 15th GMRR defensive disposition?

Indicator:

- 3 x MRCs with a total of 8 to 10 BMP-2's, and to 3 T-64B's per MRC, all in prepared fighting position or in assembly area.

194

- Main obstacle array from 800 meters to 1,000 meters forward of the MRC prepared fighting positions.

- 8 to 12 T-64B's in an assembly area.

- 2 to 3 BMP-2's forward

- 2 to 3 BMP-2's forward 1 to 3 kilometers of main defensive position with possible protective type obstacle.

- 1 BMP-2 or BMP-1 BRDM forward and isolated from any additional forces.

- 5 to 7 BRDM-2's, mounting 5 AT-5 Spandrel AT guided missiles in assembly area, possible mine layer with 1 BTR included.

PIR: Is there a weakness in the 15th GMRR defensive disposition? If so, where?

Indicator:

- Distance between MRCs greater than 2,000 meters.

- No impeding type obstacle within the main AA.

- Location of tanks within MRC positions.

- No tanks with second echelon forces.

- Location of MRR reserve.

PIR: Where are the artillery battalions comprising the 15th GMRR regimental artillery group? This PIR will also serve as an RII to higher headquarters and will facilitate the integration of ECM support by the IEWSE officer to support the scheme of maneuver.

INDICATOR: 3 x 5 to 8 2S1's or 2S3's on line, located off a major AA or MC.

PIR: Will the 41st GMRD commander direct any of their gunships against 1st brigade?

If so, where? This PIR will serve as an RII to higher headquarters.

PIR: Where and when will the 35th TR be committed against 1st brigade? This PIR will also serve as an RII to higher headquarters for initial acquisition and tracking.

INDICATOR: 60 to 70 T-64B's moving southeast from NAI 90 to NAI 16 and NAI 18.

PIR: Will the enemy use chemical munitions against 1st brigade? If so, when? This PIR will serve as an RII to higher headquarters for initial indication of intent to employ.

Indicator:

- Break off activities of enemy forces in contact.
- Enemy troops wearing protective overgarment.

Development of Situation Template and Event Template

Concurrently with the development of the indicators, CPT Roberts began to identify NAI that, when defined by the indicators, would form the basis and focus of the brigade R&S efforts. CPT Roberts' event analysis process was developed to ascertain the defensive COA as it relates to the reconnaissance and security echelon and the 2d echelon motorized rifle battalion (MRB) and MRR reserve. This is outlined in the brigade SOP. The first echelon MRBs defensive COA was given to the task forces to develop, as assisted by the brigade S2s enemy situation template and event template. Figure B-2 shows an event template. Figure B-3 is a combined situation template and event template.

Specific R&S Guidance

When MAJ Baker returned from the meeting he had the tentative brigade plan to accomplish the mission. He informed CPT Roberts of the brigade's scheme of maneuver and provided him additional R&S instructions based on the commander's concept of operations. COL Gayagas wanted a good reconnaissance of Axis Speed and Axis Kill (Annex A of the OPORD), and he wanted two OPs established: one overmatching OBJECTIVE CAT and the other OBJECTIVE DOG.

MAJ Baker told CPT Roberts that since he now knew the brigadets scheme of maneuver, he had sufficient information to assign specific informational requirements to subordinates and attached units and complete the brigade R&S plan. CPT Roberts was also to develop the R&S tasking matrix, which is the tool used to disseminate R&S taskings to subordinate and attached units. He is to have the matrix ready within the hour so MAJ Baker could pass it through COL Gayagas for his approval and MAJ Booth for coordination.

The following samples are the results of the brigade's mission analysis and decision-making process to develop the R&S plan. The sample is given along with the figure number assigned to it.

■   Figure B-4. Sample warning order.

■   Figure B-5. Sample OPORD.

■   Figure B-6. Sample Annex A to OPORD I-XX.

■   Figure B-7. Sample Annex B to OPORD I-XX.

■   Figure B-8. A sample Appendix 3 to Appendix B to OPORD 1-XX.

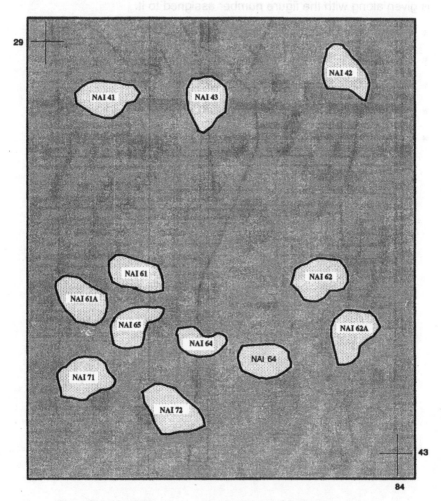

Figure B-2. Modified event template based on MRR situation template.

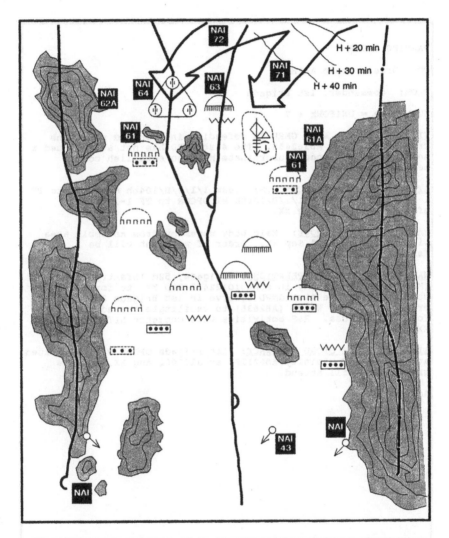

Figure B-3. Situation template and event template combined to form the foundation for the R&S plan.

```
WARNING ORDER

TO:  Orders Group ALPHA

FROM:  Commander, 1st Brigade

ZULU Time = UNIFORM + 7

SITUATION:  The 41st GMRD is defending in prepared positions
from AB448098 to AB362456.  The 41st GMRD has not established a
security zone; expect subordinate MRRs to establish combat
security outposts.

ATTACHMENTS AND DETACHMENTS:  Team 1/1/2/B/104th MI OPCON to TF
1-10 (Mech); Team 2/1/2/B/104th MI OPCON to TF 1-12(Mech)
effective 310500Z Aug XX.

EARLIEST TIME OF MOVE:  Main body movement from assembly area
will be NET 012300Z Sep XX.  Order of movement will be
designated at OPORD.

NATURE AND TIME OF OPERATIONS:  Brigade, 52d Infantry Division
(Mech) destroys enemy in zone (012245Z Sep XX) to force the
commitment of the 41st GMRD reserve in 1st Brigade Zone.  On
order, seize OBJ Watch (AB2635) to facilitate the advance of
follow-on forces.  R&S activities will commence NLT 311701Z Aug
XX.

TIME AND PLACE OPORD ISSUANCE:  1st Brigade OPORD will be issued
at brigade TOC vicinity AB621180 at 310700Z Aug XX.  Orders
group ALPHA will attend.

Acknowledge.
```

Figure B-4. Sample warning order.

```
                                      Copy ___ of ___ Copies
                                      1st Bde, 52d ID (M)
                                      (Location)
                                      310700Z Aug XX
```

OPORD 1-XX

Reference: Omitted

Time Zone Used Throughout Order: ZULU (UNIFORM + 7).

Task Organization: Omitted

1. SITUATION.

    a. Enemy Forces: Annex B (Intelligence).

    b. Friendly Forces:

        (1) 52d ID (Mech). Attacks to destroy the 41st GMRD
and seize defensible terrain from AB255385 to AB145135 NLT
012245Z Sep XX. On order, defend in sector to destroy the 9th
CAA. The Div Cdr intent is to force the commitment of the 41st
GMRD reserve into 1st Brigade Zone to allow the division main
effort in the north to quickly penetrate deep into the MRD
sector and rapidly destroy the enemy; and allow the Div more
time to prepare the defense to destroy the 9th CAA.

        (2) 2d Brigade. Division main effort in the north....

        (3) 3d Brigade. Follow and support 2d Brigade....

2. MISSION.

    1st Brigade, 52d ID (Mech) attacks 012245Z Sep XX to seize
OBJ CAT (AB4014) and OBJ DOG (AB4011) to destroy enemy in zone
and force the commitment of 41st GMRD reserve into 1st Brigade
Zone. On order, continue the attack west to seize OBJ WATCH
(AB2635). Be prepared to conduct a hasty defense on OBJ CAT and
DOG to destroy the 41st GMRD's counterattack.

3. EXECUTION.

    a. Concept of Operations. Annex A of the OPORD. I see
this as a multiphase operation. First, the reconnaissance
phase: I want an aggressive R&S effort conducted to ascertain
the disposition of the 15th GMRR. I want to find their
```

Figure B-5. Sample OPORD.
```

obstacles, to be able to take advantage of any weaknesses and avoid his strength. I need R&S assets deep to track the 41st GMRD reserve (the 35th TR, counterattack force) once it is committed in my zone. The second phase is from the LD to the objectives (CAT and DOG). TF 1-10 will be the main attack on Axis Speed; TF 1-12 will conduct the supporting attack in zone. TF 3-68 will be the Bde reserve. I want TF 1-12 to cross the LD at 012245Z Sep XX and the rest of the Bde crossing 15 minutes later. TF 1-12 will draw the 15th GMRR reserve to counterattack into its zone to facilitate TF 1-10 efforts to unhinge the enemy defense from the north quickly. We will clear enemy in zone; no enemy force platoon size or larger will be bypassed. I will interdict enemy disposition with the fire support assets then destroy them with overwhelming combat power. We cannot allow the enemy to slow our attack. The third phase will be the destruction of the 41st GMRD reserve. Once TF 1-10 and TF 1-12 seize OBJ CAT and DOG, and the 41st GMRD initiates its counterattack, TF 3-68 will pass through TF 1-10 and counterattack into the flank of the 41st GMRD reserve and destroy the enemy. The last phase is to continue the attack west to facilitate the Bde's follow-on defensive mission.

      (1) Maneuver. 1st Brigade conducts this mission in four phases.

           (a) Phase 1: Reconnaissance and surveillance.

           (b) Phase 2: TF 1-10, initial main effort; attacks on Axis Speed at 072300Z Sep XX to seize OBJ CAT (AB4014), to destroy enemy in defensive position; occupies Attack Position #1 NLT 012200Z Sep XX; LD is PL Space. TF 3-68 follows TF 1-10 as the Bde reserve. TF 1-12 occupies Attack Position #2 NLT 012200Z Sep XX, conducts supporting attack at 012245Z Sep XX, to destroy enemy in zone and seize OBJ DOG, in an effort to force the commitment of the 15th GMRR reserve into OBJ DOG (AB4011) to support TF 1-10 attack on OBJ DOG.

           (c) Phase 3: On order, TF 3-68 becomes the main effort and counterattacks on Axis Kill to destroy 41st GMRD reserve (35th TK Regt (-)); TF 1-10 supports TF 3-68 by fire, to fix enemy in engagement area RED (AB395165). TF 1-12, be prepared to support TF 3-68 and TF 1-10; on order, TF 1-12 becomes the Bde reserve.

           (d) Phase 4: 1st Bde conducts a deliberate defense on OBJ WATCH NLT 022400Z Sep XX, 1st Bde defends in sector with two TFs abreast--TF 1-10 (North) and TF 1-12 (South)--and TF 3-68 in reserve.

Figure B-5. Sample OPORD (continued).

```
     (2)  Fires.  See Annex-- (omitted)

     (3)  Obstacles, Mines, and Fortification.  See
Annex--Omitted

     (4)  IEW.  I want an aggressive R&S effort.  My first
requirement is to have a good description of the enemy's
offensive posture and find out if he is deploying any security
forces forward.  The S2 will coordinate and supervise the
effort.  The S3 will integrate my EW requirements through the
IEWSE.  I want EW support in locating the 41st GMRD, and 15th
GMRR reserves.  I want jamming support directed against the 15th
GMRR artillery resource and integrated into the Bde's scheme of
maneuver to support our movement onto the OBJs.  Additionally, I
need jamming support directed against C² nodes between the MRB
in the south and the 15th GMRR.  This will support their belief
that my main effort will be in the south.

  b.  Subordinate Units Subparagraphs.

     (1)  Combat Arms Unit.  Omitted

     (2)  Fire Support.  Omitted

     (3)  Air Defense, Aviation, Engineer, MI.  Omitted

     (4)  Reserve.  Omitted

     (5)  Coordinating Instruction.

          (a)  PIR Phase 1:

              1  What is the 15th GMRR defensive disposition
and obstacle array?

              2  What is the identification and location of
the artillery battalions comprising the 15th GMRR regimental
artillery group?

              3  Where is the 15th GMRR reserve located?

          (b)  PIR Phase 2:

              1  What is the identification and location of
the artillery battalions comprising the 15th GMRR regimental
artillery group?
```

Figure B-5. Sample OPORD (continued).

<u>2</u> Where and when is the 15th GMRR reserve being committed?

<u>3</u> Where is the 35th TR located?

<u>4</u> Will the enemy employ chemical munitions against 1st Bde? If so, when?

(c) PIR Phase 3:

<u>1</u> Where and when is the 35th TR being committed?

<u>2</u> Will the enemy employ any of their gun ships against 1st Bde? If so, when?

<u>3</u> Will the enemy employ chemical munitions against 1 Bde? If so, when?

(d) TF's R&S plans: Required at Brigade by 311400Z Sep XX.

4. **SERVICE SUPPORT.** Omitted

5. **COMMAND AND SIGNAL.** Omitted

Acknowledge.

GAYAGAS
COL

OFFICIAL

(SIGNED)
S3

Annexes: A - Operations Overlay.
          B - Intelligence:
             APP 1--Enemy Situation Overlay (Figure B-1).
             APP 2--Event and R&S Overlay (Figure B-2).
             APP 3--R&S Tasking Matrix.

Figure B-5. Sample OPORD (continued).

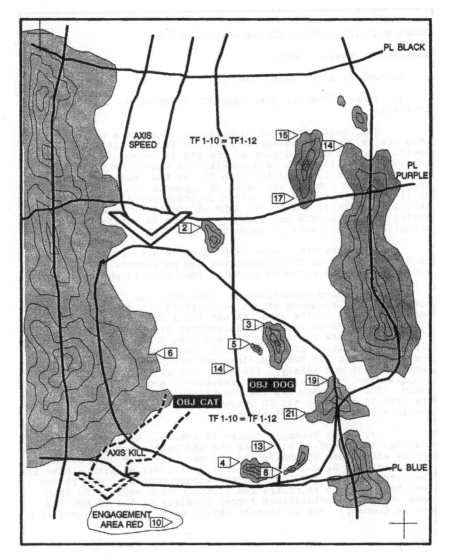

Figure B-6. Sample Annex A to OPORD 1-XX.

ANNEX B (INTELLIGENCE) TO 1st BDE OPORD 1-XX

Reference: Basic OPORD.

1. SUMMARY OF ENEMY SITUATION.

    a. Enemy Situation (See Appendix 1, Enemy Situation Overlay).

        (1) Composition or Disposition. Elements of the 41st GMRD are in prepared defensive position from AB373856 to AB448098. 41st GMRD is part of the 8th CAA's second defensive echelon. The 41st GMRD is dispositioned with 3 MRRs (15 G, 37 G, and 60 G) on line and the 35th TR, as the reserve. Committed forces against 1st Bde are shown on Appendix 1. Reinforcement is limited to the 35th TR. Known location for 41st GMRD's artillery battalions are 15th GMRR at AB3812; 37th GMRR at AB3720; and 60th GMRR at AB3635. The 37th TR artillery battalion is currently in support of the 41st GMRD DAG.

        (2) Strength. 41st GMRD elements are presently at 60 to 70 percent strength in equipment and personnel. 41st GMRD's reconnaissance is believed to be at less than 50 percent.

        (3) Enemy Capabilities. 41th GMRD and the 8th CAA are not capable of resuming offensive operations. The 385th Fighter Bomber Regiment of the 37th TAA can fly 1 to 2 sorties of MIG 27 Flogger Ds or SU-25 Frogfoot, in support of the 8th CAA. The 41st GMRD has the capability of employing 2 battalions of 122 mm (2S1), 2 battalions of 152 mm (2S3), and 3 batteries of 120 mm mortars against 1st Bde. Nuclear and biological attacks are possible, but not likely. The enemy has used persistent and nonpersistent chemical agents previously, and is likely to do so again. The 41st GMRD has the capability to counterattack with the 37th TR to blunt 52d ID (Mech) attack. Units at all levels should expect local counterattacks.

        (4) Most Probable Course of Action. The 41st GMRD will continue to defend to allow the 9th CAA sufficient time to deploy and assume offensive operations. The 15th GMRR defending within the 1st Brigade zone of attack will defend with 2 MRBs in the first echelon and 1 MRB in the second echelon. The TB has been broken up to reinforce the first echelon MRBs. The 15th GMRR reserve (counterattack force) consists of the TB HQs and one TK Company. The AT Missile (AT-5) Battery will be used as a

Figure B-7. Sample Annex B to OPORD 1-XX.

mobile reserve, with the possibility of one of its platoons sent to reinforce one of the first echelon MRBs. Although, there is no MRD-level security zone, expect combat security outposts forward of each first echelon MRB. Expect enemy artillery fire to commence as soon as the brigade moves across the LD. Enemy will use nonpersistent chemical munitions mixed in with their initial artillery fires, to slow and disrupt the brigade movement. Enemy use and intent for their obstacles will be to channelize the Bde and impede our movement within their established kill zones and fire sacks.

2. INTELLIGENCE REQUIREMENTS.

    a. PIR: see Coordinating Instruction, basic order.

    b. IR:

        (1) Has the 15th GMRR deployed combat security outposts? If so, where and what size?

        (2) Where is the 15th GMRR and AT missile battery?

        (3) Where are the 15th GMRR obstacles?

3. INTELLIGENCE ACQUISITION TASKS.

    a. Subordinate and Attached Units: See Appendix 3.

    b. Request to Higher, Adjacent, and Cooperating Units:

        (1) Will the 41st GMRD employ NBC attacks?

        (2) What enemy REC locations are in 1st Brigade zone?

        (3) When will the 37th TR be committed in 1st Brigade zone?

4. MEASURES FOR HANDLING PERSONNEL, DOCUMENTS, AND MATERIALS.

5. DOCUMENTS AND/OR EQUIPMENT REQUIRED. Omitted.

6. CI AND OPSEC. Commanders will emphasize dispersion, camouflage, noise and light discipline, and COMSEC.

Figure B-7. Sample Annex B to OPORD 1-XX (continued).

7. **REPORTS AND DISTRIBUTION.** Subordinate Units R&S Plans – submit NLT311400Z Aug XX.

8. **MISCELLANEOUS INSTRUCTION.** Omitted.

APPENDIXES:

1 – Enemy Situation Overlay

2 – Event and R&S Overlay

3 – R&S Tasking Matrix

Figure B-7. Sample Annex B to OPORD 1-XX (continued).

(CLASSIFICATION)					
R&S TASKING MATRIX					
UNIT TASKING	PRI-ORITY	NAI	LOCATION	REPORTING REQUIREMENT	REMARKS
				EVENT AND INDICATOR	
TASK FORCE 1-10					
1. Recon Axis Speed.	1		See R&S overlay	Conditions that affect trafficability and maneuverability. Obstacle type, size, and orientation.	Report as obtained.
2. Establish OP.		67	AB427185	Surveillance of activities on OBJ CAT.	Establish position NLT 311900 AUG XX.
3. Recon.		61	AB410150	3 MRP's with 8 to 10 x BMP-2s, in prepared fighting positions: Obstacle array 800 to 1,000 m forward of MRC position.	Report all occupied or unoccupied fighting positions. Obstacles: report type, size, and orientation NLT 010200Z Sep XX.
4. Recon		61A	AB400163	Alternate position for MRC at NAI 8.	
5. Recon		63	AB390133	3 MRPs with 8 to 10 x BMP2s, prepared fighting positions: Obstacle array 800 to 1,000 m forward of MRC position.	Report all occupied or unoccupied fighting positions. Obstacles: report type, size, and orientation NLT 010200Z Sep XX.
6. Recon		65	AB390150	8 to 12 x T64B's in assembly area.	Possible MRR reserve. Report NLT 010200Z Sep XX.
(CLASSIFICATION)					

Figure B-8. Sample Appendix 3 to Appendix B to OPORD 1-XX.

(CLASSIFICATION)

R&S TASKING MATRIX

UNIT TASKING	PRI-ORITY	NAI	LOCATION	REPORTING REQUIREMENT / EVENT AND INDICATOR	REMARKS
TASK FORCE 1-12					
1. Recon.	1	43	AB501098	1 BRDM or 1 BMP-2.	MRR recon company element. Report as obtained.
2. Establish OP.		60	AB416099	Surveillance of activities on OBJ DOG.	Establish positions NLT 311900 2 Aug XX.
3. Recon.		62	AB408105	3 MRP's with 8 to 10 EMP-2s, in prepared fighting positions: Obstacle array 800 to 1,000 m forward or MRC position.	Report all occupied or unoccupied fighting positions. Obstacles: report type, size, and orientation NLT 010200Z Sep XX.
4. Recon.		62A	AB384098	Alternate positions for MRC at NAI 8.	
5. Recon.		64	AB379118	3 batteries with 4 to 6 2S1 x 2S3s on line, located off a major avenue of approach or MC.	Report all positions NLT 010200Z Sep XX.
TASK FORCE 3-60					
Establish flank screen oriented north, northwest.			AB360160 to AB413370	60 to 70 x T64B's moving southeast from NAI 16 (AB380192) and NAI 18 (AD362184) heading towards NAI 10 (AB370162) or NAI 20 (AB360125).	Coordinate passage or flank screen element with TF 1-10. Establish screen NLT 011000Z Sep XX.

(CLASSIFICATION)

Figure B-8. Sample Appendix 3 to Appendix B to OPORD 1-XX (continued).

210

Glossary

AA	avenue of approach
AAA	antiaircraft artillery
ACR	armored cavalry regiment
ACRV	a type Soviet equipment
ADA	air defense artillery
AE	aerial exploitation
AF	Air Force
AI	area of interest
ALO	air liaison officer
AM	amplitude modulated
ammo	ammunition
AO	area of operations
APC	armored personnel carrier
approx	approximately
arty	artillery
ASC	armored scout car
ASPS	all-source production section
AT	antitank
attn	attention
AVLB	armored vehicle launched bridge
avn	aviation
AZ	Arizona
BAE	battlefield area evaluation
bde	brigade
BE	basic encyclopedia
bldg	building
BLUFOR	Blue Force (US)
BMNT	beginning of morning nautical twilight
BMP	a type of Soviet personnel carrier
bn	battalion
BOS	battlefield operating systems
BRDM	a type of Soviet equipment
BRM	a type of Soviet equipment
BTR	a type of Soviet equipment
CAA	combined arms army
C²	command and control

C³	command, control, and communications
C³ CM	command, control, and communications countermeasures
C&J	collection and jamming
CANE	Combined Arms in a Nuclear/Chemical Environment
CAS	close air support
cdr	commander
C-E	Communications-Electronics
CESO	communications-electronic staff officer
chem	chemical
CI	counterintelligence
CM&D	collection management and dissemination
C/MOB	countermobility
co	company
COA	course of action
Coil	collection
Comm	communication
COMSEC	communications security
con	contact
counter-reconnaissance	all measures taken to prevent hostile observation of a force, area, or place (DOD)
CP	command post
CPT	captain
CR	counterreconnaissance
CSS	combat service support
CUCV	commercial utility cargo vehicle
DA	Department of the Army
DAG	division artillery group (Soviet)
DC	District of Columbia
DD	Department of Defense
decon	decontamination
def	defense/defensive
det	detachment
DF	direction finding
DIA	Defense Intelligence Agency
dist	distribution
div	division
DIVARTY	division artillery

DS	direct support
DST	decision support template
DTG	date-time group
DZ	drop zone
E	east
ECCM	electronic counter-countermeasures
ECM	electronic countermeasures
EENT	end of evening nautical twilight
ELINT	electronic intelligence
engr	engineer
EPB	electronic preparation of the battlefield
EPW	enemy prisoner of war
ESM	electronic warfare support measures
EW	electronic warfare
FAAR	forward area alerting radar
FASCAM	family of scatterable mines
FEBA	forward edge of the battle area
FIST	fire support team
FISTV	FIST vehicle
577	tracked operations vehicle (heavy division)
FLOT	forward line of own troops
flt	flight
FM	(with number) field manual
FM	frequency modulated
FO	forward observer
FRAGO	fragmentary order
FSE	fire support element
FSO	fire support officer
fwd	forward
G2	Assistant Chief of Staff (Intelligence)
GMRD	guards motorized rifle division
GMRR	guards motorized rifle regiment
gp	group
GS	general support
GSR	ground surveillance radar
H	the time hostilities commence
HF	high frequency

HHOC	Headquarters, headquarters and operations company
HMMWV	high mobility multipurpose wheeled vehicle
HPT	high payoff target
hq	headquarters
hr	hour
HUMINT	human intelligence
HVT	high value targets
I&S	intelligence and surveillance
ID	infantry division/identification
IEW	intelligence and electronic warfare
IEWSE	intelligence and electronic warfare support element
IFF	identification, friend or foe (radar)
illum	illumination
IMINT	imagery intelligence
intcp	intercept
intel	intelligence
IPB	intelligence preparation of the battlefield
IR	information requirements
ITB	independent tank battalion
ITR	independent tank regiment
JAAT	Joint Air Attack Team
JTCG	Joint Test Command Group
km	kilometer
km/h	kilometers per hour
lbs	pounds
LC	line of contact
LD	line of departure
LIC	low-intensity conflict
LOA	limit of advance
LOC	lines of communications
LOEA	limit of enemy advance
LOS	line of sight
LP	listening post
LOB	line of bearing

m	mechanized
m	meter
mag	magnetic
MAJ	major
MC	mobility corridor
M/CM/S	mobility, countermobility, and survivability
MCOO	modified combined obstacles overlay
mech	mechanized
MEDEVAC	medical evacuation
METL	mission essential task list
METT-T	mission, enemy, terrain, troops, and time available
MI	Military Intelligence
MIJI	meaconing, intrusion, jamming, and interference
min	minute
MOPP	mission-oriented protection posture
MP	Military Police
MR	motorized rifle
MRB	motorized rifle battalion
MRC	motorized rifle company
MRD	motorized rifle division
MRP	motorized rifle platoon
MRR	motorized rifle regiment
MSR	main supply route
MTLB	a type Soviet equipment
N	north
NAI	named areas of interest
NATO	North Atlantic Treaty Organization
NBC	nuclear, biological, and chemical
NET	not earlier than
NLT	not later than
no	number
NOD	night observation device
noncomm	noncommunications
NV	night vision
NVG	night vision goggles
OB	order of battle
obj	objective
O&I	operations and intelligence

215

OMG	operational maneuver group
OP	observation post
OPCON	operational control
OPFOR	opposing force
OPLAN	operations plan
OPORD	operations order
OPSEC	operations security
PIR	priority intelligence requirements
PL	phase line
plt	platoon
PMCS	preventive maintenance checks and services
PMI	preventive maintenance inspections
POL	petroleum, oils, and lubricants
poss	possible
POV	privately owned vehicle
prep	prepare
prob	probable
R&S	reconnaissance and surveillance
RAC	reconnaissance assault company
RC	Reserve Components
REC	radio electronic combat
recon	reconnaissance
REMBASS	Remotely Monitored Battlefield Sensor System
reconnaissance	A mission undertaken to obtain, by visual observation or other detection methods, information about the activities and resources of an enemy or potential enemy; or to secure data concerning the meteorological, hydrographic, or geographic characteristics of a particular area. (DOD, NATO)
regt	regiment
retrans	retransmission
RII	request for intelligence information
RISTA	reconnaissance, intelligence, surveillance and target acquisition
Rkh	a type of Soviet equipment
Rkh/m	a type of Soviet equipment
RP	release point

216

s	south
S2	Intelligence Officer (US Army)
S3	Operations and Training Officer (US Army)
S3-air	Air Operations and Training Officer (US Army)
S-A	seismic-acoustic
SA	surface to air
SALT	size, activity, location, and time
SALUTE	size, activity, location, unit, time, equipment (spot report format)
SAM	surface to air missile
SCARF	standard collection asset request format
scty	security
SEAD	suppression of enemy air defense
SIGINT	signals intelligence
SIR	specific information requirements
SITMAP	situation map
SMS	simultaneous monitoring system
SOI	signal operation instructions
SOP	standing operating procedure
SOR	specific orders and requests
SP	start point
SSM	surface-to-surface missile
Survl	surveillance
surveillance	The systematic observation of aerospace, surface or subsurface areas, places, persons, or things by visual, aural, electronic, photographic, or other means. (DOD, NATO)
TA	theater army
TAA	tactical air army
TACFIRE	tactical fire direction computer system
TAI	target areas of interest
TB	tank battalion
TC	training circular
TE	tactical exploitation
temp	temperature
TF	task force
tk	tank
TNT	trinitrotoluene

TOC	tactical operations center
TOT	time over target
TOW	tube-launched, optically tracked, wire guided
TPL	time phase line
TR	tank regiment
TRADOC	United States Army Training and Doctrine Command
TTP	tactics, techniques, and procedures
UHF	ultra high frequency
U.S.	United States
USAF	United States Air Force
USAICS	U.S. Army Intelligence Center and School
USSR	Union of Soviet Socialist Republics
VHF	very high frequency
vic	vicinity
W	west

References

Required Publications

Required publications are sources that users must read in order to understand or to comply with this field manual.

Field Manuals (Fms)

FM 17-98	Scout Platoon. October 1987.
FM 34-1	Intelligence and Electronic Warfare Operations. July 1987.
FM 34-2	Collection Management. 22 October 1990.
FM 34-3	Intelligence Analysis. 15 March 1990.
FM 34-10-1	Tactics, Techniques, and Procedures for the Remotely Monitored Battlefield Sensor Systems (REMBASS). November 1990.
FM 34-80	Brigade and Battalion Intelligence and Electronic Warfare Operations. April 1986.
FM 34-130	Intelligence Preparation of the Battlefield. May 1989.
FM 101-5	Staff Organization and Operations. May 1984.
FM 101-5-1	Operational Terms and Symbols. October 1985.
61 JTCG/ME-87-10	Handbook for Operational Testing of Electro-Optical Systems in Battlefield Obscurants. October 1987.

Department of the Army Pamphlet (DA Pam)

DA Pam 381-3	How Latin America Insurgents Fight. June 1986.
DD Form 1975	Joint Tactical Air Reconnaissance and Surveillance Request Form.

Rand Study. Applying the National Training Center Experience: Tactical Reconnaissance. October 1987.

Related Publications

Related publications are sources of additional information. They are not required in order to understand this publication.

Field Manuals (FMs)

FM 6-20-10	TTP for the Targeting Process, Mar 90
FM 24-33	Communications Techniques: Electronic Counter-Countermeasures. 22 Mar 85
FM 34-10	Division IEW Operations. November 1986.
(U) FM 34-1OA	Division Intelligence and Electronic Warfare Operations, Secret. December 1986.
FM 100-5	Operations. May 1986.
FM 101-5	Staff Organization and Preparation. May 1984.

(U) DIA Study. Reconnaissance and Surveillance and Target Acquisition in the USSR, Secret/NOFORN. June 1988.

Index

AA   See avenue of approach.

ADA See air defense artillery.

ADA officer. See air defense
   artillery officer.

aerial fire support officer
   mission support, 57

AI See area of interest.

air AA
   enemy use of, 14

air and armored cavalry
   squadron
   mission support, 58

air defense artillery
   defensive area of interest for
   Div ADA, 99

air defense artillery officer
   R&S asset, 44
   staff responsibility, 63

air defense artillery platoon
   mission, 53

air liaison officer
   coordination, 53
   JAAT coordination, 54
   See chapter 9.
   staff responsibility, 63

all-source production section
   support to CM&D process, 4

ALO See air liaison officer.

antitank (AT) helicopter
   commanders information
   requirement, 14

AO See area of operation.

area
   R&S location of operations, 8,
   18, 36
   and 41

area of interest
   identification of, 23
   offensive and defensive, 22

area of operation
   IPB process
   requirement, 23

armor
   commanders information
   requirement, 14

armored vehicle
   commanders information
   requirement, 14

Army aviation
   digital interface mission
support, 58
   mission, 53
   R&S asset, 44

Army aviation support officer
   mission coordination, 58

artillery
   commanders information
   requirement, 14

221

ASPS See all-source production section.

asset
tasking assets, 2, 4

assets and equipment
support to R&S plans, 41

attack helicopter battalion
mission support, 58

augment
scout combat missions, 40
See chapter 8.

augmentation
See chapter 8.

augmented
essential equipment, 41
scout limitations, 40

augmenting
See chapter 8.

avenue of approach
R&S application, 23
support to commanders needs, 12
terrain factor in support to R&S application, 23

aviation officer
staff responsibility, 63

battle task
development of, 5
list of, 7
platoon 7 and 8

BAE See battlefield area evaluation.

battlefield area evaluation
component of, 19

battle damage assessment
in support of TAI, 36

battlefield operating system
part of DST process, 33

BOS See battlefield operating systems

CANE See chemical environment.

CAS See close air support.

CESO See communications-electronics staff officer.

changing double column to a diamond formation
LIC movement, 136

checkpoint
R&S overlay. 72

chemical environment
IIB Test in NBC environment, 8

chemical officer
See chapter 10.

chemical-engineer reconnaissance
See chapter 10.

CI see counterintelligence.

CI team
R&S coordination, 65

civilian population
LIC operations, 141

close air support
EW operations, 161

CM&D See collection and
dissemination.

collection and dissemination
support to R&S, 3
and 4

collection asset
tasking of, 6 and 31

collection effort
battalion forces level, 39
requirements for information,
18, 26, and 31

collection management
process
R&S and the CM&D process,
2, 3, and 4

collection plan
control of the collection
requirement, 12 and 13
phases to develop a, 3
unit collection
requirements, 4

column
formation, 134

combat
category of patrol, 41
mission to support R&S, 5,
9, and 16

combat patrol
provides, 41

communications-electronics staff
officer
Brigade operations
interface, 165

concealment and cover
R&S application, 31

counterintelligence
mission support, 55
staff responsibility, 63

counterreconnaissance
CR efforts, 2
CR operations use of NAI, 6
(see NAI)
TTP and CR, iv

CR See
counterreconnaissance.

critical task
supporting battle tasks, 6, 7
and 8

cueing
R&S planning, 64

database
IPB support for R&S
planning, 16

day and night observation
device
unit capability, 42

deceive
LIC operations, 154

decision point
support relationship
to, 36

decision points or line
part of the DST process, 33

decision support template
IPB product, 32
staff responsibility, 65

defensive and offensive areas

of interest
requirements for R&S
planning, 22

demonstration
LIC threat, 115

diamond formation
insurgent movement, 135

disseminating
PIR, 16
see chapter 5

division
organization support, 2, 3,
and 4

division reconnaissance asset
division, 100

doctrinal template
IPB product, 25
product integration, 25

double column
insurgent movement, 136

DST See decision support
template.

ECCM See electronic
counter-countermeasures.

ECM See electronic
counter-measures.

electronic counter-measures
EW asset, 153

electronic
counter-countermeasures
EW asset, 153

electronic HPT
EW planning, 157

electronic support
measures
collection, 154

electronic warfare
CR battle planning and, 12

electronic warfare collection
system
mission support, 55

enemy order of battle
data base
requirement, 16

enemy prisoner of war
interrogator
mission, 55

enemy rates of
advance
See threat evaluation.

enemy situation template
IPB product, 4

engineer
augmentation support to
R&S, 36

engineer and air defense
artillery
IPB product, 22
mission support, 56

engineer officer
staff responsibility, 64

engineer platoon
mission, 52
R&S asset, 52

engineer product
IPB special product, 18

engineer section 64

engineer support officer 52

environmental effects for R&S
see Figure 25

EPW interrogator
See enemy prisoner of war
interrogator.

equipment
R&S asset, 39

ESM See electronic support
measures.

event analysis matrix
IPB product for specific
events, 33

event template
IPB product, 4 and 33
process for, 31

EW See electronic warfare.

fan
insurgent movement, 134

fan formation
insurgent movement, 137

FEBA See forward edge of the
battle area.

field artillery
integration into R&S and CR
plans, 52
mission support, 56
R&S asset, 44

field of fire
R&S application, 24

fire support
see fire support officer to
R&S, 36

fire support officer
in R&S planning, 5
JAAT coordination, 54
R&S planning, 5
staff coordination, 2
staff responsibility, 63

fire support team
mission, 52 and 54

FIST See fire support team.

FLOT See forward line of own
troops.

FO See forward observers.

forward edge of the battle
area
enemy movement
to, 31

forward line of own troops
enemy distance from, 20

forward observers
mission, 52 and 53

FRAGO See fragmentary
order.

fragmentary order
method of tasking, 67

FSO See fire support officer.

G2
division, 56
tasking, 4

GSR See ground surveillance
radar.

GSR and REMBASS
mission support 55

ground surveillance radar
24-hour coverage, 44

mission, 44
R&S asset, 44

GUARDRAIL
collection, 150

guerrilla
LIC threat, 115
support requirements, 116

guerrilla platoon
mission, 122

guerrilla training complex 128

helicopter
LIC support, 150

high payoff target
developing, 2-2
monitoring, 77

high value target
supporting the identification of
HPT's, 26

HPT See high payoff target.

HVT See high value target.

IEW See intelligence and
electronic warfare.

IEWSE See intelligence and
electronic warfare support
element.

IEWSE officer
staff responsibility, 63

IIB Test See CANE.

incident matrix
overlay support, 146
see figure 11-23

incident overlay
historical data, 141

incursion
operations, 117

indicator
effective R&S planning, 61

infantry 40

infantry platoon 89

infiltration
operations, 117

insurgent
conduct, 116
LIC threat, 115

intelligence acquisition task
method of tasking, 67

intelligence and electronic warfare
system support to
R&S, 2

intelligence and electronic
warfare support element
coordination with
R&S, 2
mission, 53

intelligence annex
use of, 16

intelligence cycle
crucial phase in R&S, 12

intelligence preparation of the
battlefield
R&S and IPB, 2
and 5

intelligence requirement
commanders needs, 3,

(see PIR and R&S)

intelligence summaries
use of, 16 (see intelligence
annex)

intercept
IEW asset, 155

interrogation
LIC support, 148

IPB See intelligence preparation
of the battlefield.

IR See intelligence
requirement.

JAAT See Joint Air Attack
Team.

jam
LIC operations, 155

jammer
brigade operations, 166

Joint Air Attack Team
OH-58D interface, 54

key terrain
R&S application, 23

"L" formation
insurgent movement, 134

LIC See chapter 11.

light infantry See chapter 10.

limit of enemy advance
identify, 107

limits of responsibility
in the R&S effort, 37

listening post
enemy, 15

maneuver element, 41
R&S asset, 39

locate
LIC operations, 155

LOEA See limit of enemy
advance.

LP See listening post.

matrix
method, 73

MCOO See modified combined
obstacles overlay.

mechanized infantry
see chapter 10

METT-T See mission, enemy,
terrain, time, and troops
available.

military police platoon
mission support, 59

mission analysis
AI determination, 19

mission, enemy, terrain, time, and
troops available
unit responsibilities to R&S
planning, 37

modified combined obstacles
overlay, 70

movement formation
insurgent operations, 134

MP See military police platoon.

named areas of interest
part of situation templating, 31
GSR task, 44
use of 6 (see CR)

NAI See named areas of interest.

NBC See nuclear, biological, and chemical.

NBC officer
staff responsibility,

night observation device
equipment, 42

night vision device 47

NOD See night observation device.

nuclear, biological, and chemical environment, 6, 7, and 8

OB See order of battle.

observation
R&S application, 23

observation equipment
used by scouts, 41

observation equipment associated with maneuver battalion 44

observation post
enemy, 14
maneuver element, 42
scout operations, 40

obstacle
R&S application, 23

obstacle analysis
provided by, 53

offensive operation
see chapter 9

OP See observation post.

operations order See chapter 5.

operations plan 67

OPLAN See operations plan.

OPORD See operations order.

order of battle 16

overlay
method, 73

patrol
maneuver element, 41
R&S scout, 39

PIR See priority intelligence requirement.

population status overlay
LIC support, 146

priority intelligence requirement
commanders needs, 3 (see IR and R&S)

QUICKFIX
R&S asset, 150

R&S See reconnaissance and surveillance.

R&S tasking See chapter 5.

R&S limit of responsibility See Figure 2-15.

R&S overlay
dissemination of R&S, 2
overlay, 71

R&S tasking matrix See

figure 80

radio intercept
R&S asset, 148

raid
mission, 128

reconnaissance
active, 11
as a patrol, 40
attack outcome and, 1
TTP and, iv

reconnaissance and
surveillance
develop solutions in R&S
plans, 6
importance of R&S, 1
R&S and PIR and IR, 3
(see PIR and IR)
(see scheme of
maneuver, 5)
solutions to common errors in
planning, 4

reconnaissance company See
chapter 10.

reconnaissance battalion
acquisition See chapter 10.

reconnaissance subsquad
organization, 120

redundancy
R&S planning, 64
and 65

regiment
reconnaissance unit, 101

REMBASS See remotely
monitored battlefield sensor
system.

remotely monitored battlefield
sensor system
capabilities, 48
R&S asset, 44
teams and equipment, 48
and 49

release point
R&S overlay, 72

request for intelligence
information
information needs, 16

retask
R&S assets, 85

RII See request for intelligence
information.
RP See release point.

route
task, 8
type of patrol mission, 41

route reconnaissance
scout mission, 40
scout task, 8

S2
role of S2 in planning of
R&S, iv
S2 coordination with S3 and
FSO, 2
S2 unit role, 1

S3
JAAT coordination, 54
S3 coordination with S2 and
FSO, 2
S3 involvement with
subordinate commands, 5

S3-air
  JAAT coordination, 53

sabotage
  operations, 117
  subsquads, 120

scheme of maneuver
  formulation of R&S plan, 5
  (see R&S)

scout
  R&S asset, 39

scout platoon
  operations, 40
  platoon tasks, 5, 6, 7,
  and 8

scouts with infantry
  mission require-
  ments, 40
  See chapter 8.

screening
  scout mission, 2
  and 40

security
  combat patrol capability, 41
  and 42
  patrol requirement, 41
  scout mission, 40
  scout platoon task, 7
  and 8

sensor
  LP/OP 24-hour
  capability, 42

seven BOS See chapter 9.

single
  formation, 134

single file or firing line
  formation

insurgent movement, 135

SIR See specific information
  requirement.

situation map
  insurgents, 143

situation template
  IPB product, 4
  product integration, 26

soldier
  information asset, 9
  and 42

SOR See specific orders and
  requests.

SP See start point.

special uses and effects of
  terrain
  for R&S planning, 13

specific information
  requirement
  R&S planning, 61 and 108

specific orders and request
  R&S planning, 61

start point
  R&S overlay, 72

subordinate unit instruction
  method of tasking, 67

subsquad
  guerrilla units, 120

surprise attack
  operations, 117

surveillance
  passive, 11

synchronization matrix
  part of the DST process, 33

synchronization See synchronize.

synchronize
battlefield operating systems, 33
R&S with battle, 16
the R&S effort, 36

tactical Air Force
mission, 54
R&S asset, 44

TAI See target area of interest.

target
enemy, 26
R&S planning, 4
REMBASS identification of, 48, 49, and 52

target acquisition
fire finder radar, 56
R&S and CR product require ment, 26

target acquisition asset
55

target acquisition data
provided by GSR, 44

target areas of interest
interdicting sites, 36
part of DST process, 33

target identification
48

target list worksheet
158

target value analysis
process
to determine HVT's, 26

target-acquisition

FISTV capability, 52
OH-58D capability, 53
to determine HVT's, 26

targeting
part of a successful defense, 1

targeting data
a collection requirement, 12

targeting process
36

task organize
increase scout
capabilities, 40

task organized See chapter 8.

task organizing
R&S planning, 64
See chapter 8.

tasking See chapter 5.

tasking asset
requirement, 84

terrain
analysis provided by, 52
part of R&S planning, 3 and 15

terrain analysis
component of IPB, 23

terrorist act
LIC threat, 115

threat database
information for IPB, 26

threat evaluation
component of IPB, 8, 25, and 26

threat integration

component of IPB, 8
and 26

time phase line
part of the DST
process, 31

TPL See time phase line.

trap map
insurgent, 145

two-echelon formation
insurgent movement, 137

war gaming
DST a product of, 82

war gaming process
the process of war gaming, 108

weather analysis
component of IPB, 23

weather
required for R&S planning, 3
and 15

wedge formation
insurgent movement, 134

vegetation See chapter 2.

zone
R&S patrol locations, 41
scout R&S area coverage, 40

zone reconnaissance
scout mission, 40
scout task, 8